Rachel Harriette Busk

The Valleys of Tirol

Rachel Harriette Busk

The Valleys of Tirol

ISBN/EAN: 9783741149184

Manufactured in Europe, USA, Canada, Australia, Japa

Cover: Foto ©berggeist007 / pixelio.de

Manufactured and distributed by brebook publishing software (www.brebook.com)

Rachel Harriette Busk

The Valleys of Tirol

THE
VALLEYS OF TIROL

THE
VALLEYS OF TIROL

THEIR TRADITIONS AND CUSTOMS
AND HOW TO VISIT THEM

BY

MISS R. H. BUSK

AUTHOR OF 'PATRAÑAS' 'SAGAS FROM THE FAR EAST'
'FOLK-LORE OF ROME' ETC.

WITH FRONTISPIECE AND THREE MAPS

LONDON
LONGMANS, GREEN, AND CO.
1874

PREFACE.

THERE are none who know TIROL but are forward to express regret that so picturesque and so primitive a country should be as yet, comparatively with other tracks of travel, so little opened up to the dilettante explorer.

It is quite true, on the other hand, that just in proportion as a country becomes better known, it loses, little by little, its merit of being primitive and even picturesque. Intercourse with the world beyond the mountains naturally sweeps away the idiosyncracies of the mountaineers; and though the trail of progress which the civilized tourist leaves behind him cannot absolutely obliterate the actual configuration of the country, yet its original characteristics must inevitably be modified by the changes which his visits almost insensibly occasion. The new traditions which he brings with him of vast manufacturing enterprise and rapid commercial success cannot but replace in the minds of the people the old traditions of the fire-side and the *Filò*, with their dreams of treasure-granting dwarfs and the *Bergsegen* dependent on prayer. The uniform erections of a monster Hotel Company, 'convenient to the Railway Station,' supersede the frescoed or timbered hostelry perched on high to receive the wayfarer at his weariest. The giant mill-chimneys, which sooner or later spring up from seed unwittingly scattered by the way-side, not only mar the landscape with their intrinsic deformity, but actually strip the mountains of their natural covering, and convert wooded slopes into grey and barren wastes;[1] just as the shriek of the whistle overpowers the Jödel-call, and the barrel-organ supersedes the zitther and the guitar.

Such considerations naturally make one shrink from the responsibility of taking a part (how insignificant soever) in directing the migration of tourists into such a country as Tirol. I have heard a Tirolese, while at the same time mourning that the attractions of his country were so often passed over, express this feeling very strongly, and allege it as a reason why he did not give the result of his local observations to the press; and I listened to his apprehensions with sympa-

---

[1] This is what the introduction of manufactories is doing in Italy at this moment. The director of a large establishment in Tuscany, which devours, to its own share, the growth of a whole hill-side every year, smiled at my simplicity when I expressed regret at hearing that no provision was made for replacing the timber as it is consumed.

PREFACE. vii

thy. But then these changes *must* be. The attempt to delay them is idle; nor would individual abstention from participating in the necessary movement of events have any sensible effect in stemming the even course of inevitable development. Circumstances oblige us continually to co-operate in bringing about results which we might personally deprecate.

> 'In whatsoe'er we perpetrate
> We do but row; we're steered by fate.'

And after all, why should we deprecate the result? We all admire the simple mind and chubby face of childhood; yet who (except the sentimental father in the French ballad, 'Reste toujours petit!') would wish to see his son in petticoats and leading-strings all his days. The morning mists which lend their precious charm of mystery to the sunrise landscape must be dispelled as day advances, or day would be of little use to man.

The day cannot be all morning; man's life cannot be all infancy; and we have no right so much as to wish—even though wishes avail nothing—that the minds of others should be involved in absurd illusions to which we should scorn to be thought a prey ourselves.

Nature has richly endowed Tirol with beauty and healthfulness; and they must be dull indeed who, coming in search of these qualities, do not find them enhanced an hundredfold by the clothing of poetry with which the people have superinduced them. Who, in penetrating its mountain solitudes, would not thank the guide who peoples them for him with mysterious beings of transcendent power; who interprets for him, in the nondescript echoes of evening, the utterances of a world unknown; and in the voices of the storm and of the breeze the expression of an avenging power or the whisperings of an almighty tenderness.

But then—if this is found to be something more than poetry, if the allegory which delights our fancy turns out to be a grotesque blunder in the system of the peasant who narrates it,—it cannot be fair to wish that he should continue subject to fallacious fancies, in order that we may be entertained by their recital.

It is one thing for a man who has settled the grounds of his belief (or his unbelief) to his best satisfaction in any rational way, to say, 'I take this beautiful allegory into my repertory; it elevates my moral perceptions and illustrates my higher reaches of thought;' but it is quite another thing if one reasons thus with himself, 'My belief is so and so, *because* a certain supernatural visitation proves it;' when actually the said supernatural visitation never took place at all, and was nothing but an allegory, or still less, a mere freak of fancy in its beginning.

PREFACE. ix

Perhaps if the vote could be taken, and if desires availed anything, the general consensus of thinking people would go in favour of the desire that there had been no myths, no legends. But the vote would involve the consequence that we should have antecedently to be possessed of a complete innate knowledge of the forces of being, corresponding to the correct criteria, which we flatter ourselves do indwell us of the principles of beauty and of harmony. If there are any who are sanguine enough to believe that science will one of these days give us a certain knowledge of how everything came about, it is beyond dispute that for long ages past mankind has been profoundly puzzled about the question, and it cannot be an uninteresting study to trace its gropings round and round it.

Perfect precision of ideas again would involve perfect exactness of expression. No one can fail to regret the inadequacies and vagaries of language which so often disguise instead of expressing thought, and lead to the most terrible disputes just where men seek to be most definite. If we could dedicate one articulate expression to every possible idea, we should no longer be continually called to litigate on the meanings of creeds and documents, and even verbal statements.

But when we had attained all this, we should have surrendered all the occupation of conjecture and all the charms of mystery; we should have parted with all poetry and all *jeux d'esprit*. If knowledge was so positive and language so precise that misunderstanding had no existence, then neither could we indulge in metaphor nor *égayer la matière* with any play on words. In fact, there would be nothing left to say at all!

Perhaps the price could not be too high; but in the meantime we have to deal with circumstances as they are. We cannot suppress mythology, or make it non-existent by ignoring it. It exists, and we may as well see what we can make of it, either as a study or a recreation. Conjectures and fancies surround us like thistles and roses; and as brains won't stand the wear of being ceaselessly carded with the thistles of conjecture, we may take refuge in the alternative of amusing ourselves on a holiday tour with plucking the roses which old world fancy has planted—and planted nowhere more prolifically than in Tirol.

In speaking of Tirol as comparatively little opened up, I have not overlooked the publications of pioneers who have gone before. The pages of Inglis, though both interesting and appreciative, are unhappily almost forgotten, and they only treat quite incidentally of the people's traditions. But as it is the most salient points of any matter which must always arrest attention first, it has been chiefly the mountains of Tirol to which attention has hitherto been drawn. Besides the universally useful 'Murray' and others, very efficient guidance to them has of late years been afforded in the pages of 'Ball's Central Alps,' in some of the contributions to 'Peaks, Passes, and Glaciers;' in the various works of Messrs. Gilbert and Churchill; and now Miss A. B. Edwards has shown what even ladies may do among its Untrodden Peaks. The aspects of its scenery and character, for which it is my object on the other hand to claim attention, lie hidden among its Valleys, Trodden and Untrodden. And down in its Valleys it is that its traditions dwell.[1]

If the names of the Valleys of Tirol do not at present awaken in our mind stirring memories such as cling to other European routes whither our steps are invited, ours is the fault, in that we have overlooked their history. The past has scattered liberally among them characteristic landmarks dating from every age, and far beyond the reach of dates. Every stage even of the geological formation of the country—which may almost boast of being in its courage and its probity, as it does boast of being in the shape in which it is fashioned, the heart of Europe—is sung of in popular *Sage* as the result of some poetically conceived

---

[1] Except the Legends of the Marmolata, which I have given in 'Household Stories from the Land of Hofer; or, Popular Myths of Tirol,' I hardly remember to have met any concerning its prominent heights.

agency; humdrum physical forces transformed by the wand of imagination into personal beings; now bountiful, now retributive; now loving; now terrible; but nearly always rational and just.

To the use of those who care to find such gleams of poetry thrown athwart Nature's work the following pages are dedicated. The traditions they record do not claim to have been all gathered at first hand from the stocks on which they were grown or grafted. A life, or several lives, would hardly have sufficed for the work. In Germany, unlike Italy, myths have called into being a whole race of collectors, and Tirol has an abundant share of them among her offspring. Not only have able and diligent sons devoted themselves professionally to the preservation of her traditions, but every valley nurtures appreciative minds to whom it is a delight to store them in silence, and who willingly discuss such lore with the traveller who has a taste for it.

That a foreigner should attempt to add another to these very full, if not exhaustive collections, would seem an impertinent labour of supererogation. My work, therefore, has been to collate and arrange those traditions which have been given me, or which I have found ready heaped up; to select from the exuberant mass those which, for one reason or another, appeared to possess the most considerable interest; and to localise

PREFACE.

them in such a way as to facilitate their study both by myself and others along the wayside; not neglecting, however, any opportunity that has come in my way of conversing about them with the people themselves, and so meeting them again, living, as it were, in their respective homes. This task, as far as I know, has not been performed by any native writer.[1]

The names of the collectors I have followed are, to all who know the country, the best possible guarantee of the authenticity of what they advance; and I subjoin here a list of the chief works I have either studied myself or referred to, through the medium of kind helpers in Tirol, so as not to weary the reader as well as myself with references in every chapter:—

> Von Alpenburg: Mythen und Sagen Tirols.
> Brandis: Ehrenkränzel Tirols.
> H. J. von Collin: Kaiser Max auf der Martinswand: ein Gedicht.
> Das Drama des Mittelalters in Tirol. A. Pickler.
> Hormayr: Taschenbuch für die Vaterländische Geschichte.
> Meyer: Sagenkränzlein aus Tirol.
> Nork: Die Mythologie der Volkssagen und Volksmärchen.
> Die Oswaldlegende und ihre Beziehung auf Deutscher Mythologie.
> Oswald v. Wolkenstein: Gedichte. Reprint, with introduction by Weber.
> Perini: I Castelli del Tirolo.
> Der Pilger durch Tirol; geschichtliche und topographische Beschreibung der Wallfahrtsorte u. Gnadenbilder in Tirol u. Vorarlberg.

---

[1] I published much of the matter of the following pages in the first instance in the *Monthly Packet*, and I have to thank the Editor for my present use of them.

A. Pickler: Frühlieder aus Tirol.
Scherer: Geographie und Geschichte von Tirol.
Simrock: Legenden.
Schneller: Mährchen und Sagen aus Wälsch-Tirol.
Stafler: Das Deutsche Tirol und Vorarlberg.
Die Sage von Kaiser Max auf der Martinswand.
J. Thaler: Geschichte Tirols von der Urzeit.
Der Untersberg bei Salzburg, dessen geheimnissvolle Sagen der Vorzeit, nebst Beschreibung dieses Wunderberges.
Vonbun: Sagen Vorarlbergs.
Weber: Das Land Tirol. Drei Bänder.
Zingerle: König Laurin, oder der Rosengarten in Tirol.
    Die Sagen von Margaretha der Maultasche.
    Sagen, Märchen u. Gebräuche aus Tirol.
    Der berühmte Landwirth Andreas Hofer.

I hope my little maps will convey a sufficient notion of the divisions of Tirol, the position of its valleys and of the routes through them tracked in the following pages. I have been desirous to crowd them as little as possible, and to indicate as far as may be, by the size and direction of the words, the direction and the relative importance of the valleys.

Of its four divisions the present volume is concerned with the first (Vorarlberg), the fourth (Wälsch-Tirol), and with the greater part of the valleys of the second (Nord or Deutsch-Tirol.) In the remoter recesses of them all some strange and peculiar dialects linger, which perhaps hold a mine in store for the philologist. Yet, though the belief was expressed more than thirty years ago [1] that they might serve as a key to the

---

[1] See Steub 'Über die Urbewohner Rätiens und ihren Zusammenhang mit den Etruskern. München, 1843,' quoted in *Dennis' Cities and Cemeteries of Etruria,* I. Preface, p. xlv.

Etruscan language, I believe no one has since been at the pains to pursue this most interesting research. In the hope of inducing some one to enter this field of enquiry, I will subjoin a list of some few expressions which do not carry on their face a striking resemblance to either of the main languages of the country, leaving to the better-informed to make out whence they come. The two main languages (and these will suffice the ordinary traveller for all practical purposes), are German in Vorarlberg and North Tirol, Italian in Wälsch-Tirol, mixed with occasional patches of German; and in South-Tirol with a considerable preponderance of these patches. A tendency to bring about the absorption of the Italian-speaking valleys into Italy has been much stimulated in modern times, and in the various troubled epochs of the last five-and-twenty years Garibaldian attacks have been made upon the frontier line. The population was found stedfast in its loyalty to Austria, however, and all these attempts were repulsed by the native sharp-shooters, with little assistance from the regular troops. An active club and newspaper propagandism is still going on, promoted by those who would obliterate Austria from the map of Europe. For them, there exists only German-Tirol and the Trentino. And the Trentino is now frequently spoken of as a province bordering on, instead of as in reality, a division of, Tirol.

Although German is generally spoken throughout Vorarlberg, there is a mixture of Italian expressions in the language of the people, which does not occur at all in North-Tirol: as

> *fazanedle*, for a handkerchief (Ital. *fazzoletto*.)
> *gaude*, gladness (Ital. *gaudio*.)
> *guttera*, a bottle (Ital. *gutto* a cruet.)
> *gespusa*, a bride (Ital. *sposa*).
> *gouter*, a counterpane (Ital. *coltre*).
> *schapel*, the hat (peculiar to local costume), (Ital. *cappello*, a hat).

The *k* in many German words is here written with *ch*; and no doubt such names as the Walgau, Walserthal, &c., commemorate periods of Venetian rule.

Now for some of the more 'outlandish' words:—

> *baschga'* (the final *n*, *en*, *rn*, &c. of the German form of the infinitive is usually clipped by the Vorarlbergers, even in German words, just as the Italians constantly clip the final letters of their infinitive, as *anda'* and *andar'* for *andare*, to walk, &c.) to overcome.
> *batta'*, to serve.
> *pütze'* or *buetza'*, to sew or to piece.
> *häss*, clothing.
> *res*, speech.
> *tobel*, a ravine.
> *fed*, a girl; *spudel*, an active girl; *schmel*, a smiling girl.
> *hattel*, a goat; *mütl*, a kid.
> *Atti*,[1] father, and *datti*, 'daddy.'
> *frei*, pleasant.
> *zoana*, a wattled basket.
> *schlutta* and *schoope*, a smock-frock.
> *täibe*, anger.

---

[1] See it in use below, p. 28, and comp. *Etruscan Res.* p. 302, note.

kiba', to strive.
rûra' to weep.[1]
musper, merry.
tribiliera', to constrain.
waedle, swift.
raetig werden, to deliberate.
Tripstrüll, = Utopia.
weeh, spruce, also vain.
laegla, a little vessel.
hengest, a friendly gathering of men.[2]
koga, cursed, also corrupted.
fegga, a wing.
krom, a gift.
blaetz, a patch.
grind, a brute's head, a jolterhead.
bratza, a paw, an ugly hand.
briegga', to pucker up the face ready for crying.
deihja, a shepherd's or cattle-herd's hut.[3]
also *dieja*, which is generally reserved for a hut formed by taking advantage of a natural hole, leaving only a roof to be supplied.
garreg, prominent. (I *think* that *gareggiante* in Italian is sometimes used in a similar sense.)

Other words in Vorarlberg dialect are very like English, as:—

Witsch, a witch.
Pfülle, a pillow.

---

[1] Somewhat like *pleurer*. A good many words are like French, as *gutschle*, a settle (couche); *schesa*, a gig; and *gespusa*, mentioned above, is like épouse; and *au*, for water, is common over N. Tirol, as well as Vorarlberg, *e.g. infra*, pp. 24, 111. &c.

[2] Comp. *Etrus. Res.* 339-41.

[3] Several places have received their name from having grown round such a hut; some of these occur outside Vorarlberg, as for instance Kühthei near St. Sigismund (*infra*, p. 331) in the Lisenthal, and Niederthei in the Œtzthal.

> *rôt*, wrath.
> *gompa'*, to jump.
> *gülla*, a gulley.
> also *datti* and *schmel*, mentioned already.
> *Aftermötig* (after-Monday) is a local name for Tuesday.

In Wälsch-Tirol, they have

> *carega*, a chair.
> *bagherle*, a little carriage, a car.
> *troz*, a mountain path.
> *Malga*,[1] equivalent to Alp, a mountain pasture.
> *zufolo*,[2] a pipe.
> And *Turlulù* (*infra*, p. 432) is nearly identical in form and sound with a word expounded in *Etrus. Researches*, p. 299.
> Of 'Salvan' and 'Gannes,' I have already spoken.[3]

But all this is, I am aware, but a mere turning over of the surface; my only wish is that some one of stronger capacity will dig deeper. Of many dialects, too, I have had no opportunity of knowing anything at all. Here are, however, a few suggestive or strange words from North and South Tirol:—

Pill, which occurs in various localities[4] of both those provinces to designate a place built on a little hill or knoll, is identical with an Etruscan word to which Mr. Isaac Taylor gives a similar significance.[5] I do not overlook Weber's observation that 'Pill is obviously a corruption of *Büchel* (the German for a a knoll), through *Bühel* and *Bühl*;' but, which pro-

---

[1] Comp. *ma* = earth, land, *Etrus. Res.* pp. 121, 285.
[2] Comp. *subulo*, *Etrus. Res.* 324. Dennis i. 339.
[3] *Infra*, p. 411.
[4] See e.g., *infra*, p. 202.
[5] *Etrus. Res.* p. 330.

PREFACE.

ceeds from which is often a knotty point in questions of derivation, and Weber did not know of the Etruscan 'pil.'

Ziller and *celer* I have already alluded to,[1] though of course it may be said that the Tirolean river had its name from an already romanised Etruscan word, and does not necessarily involve direct contact with the Etruscan vocabulary.

> *Grau-wutzl* is a name in the Zillerthal for the Devil.
> *Disel*, for disease of any kind.
> *Gigl*, a sheep.
> *Kiess*, a heifer.
> *Triel*, a lip.
> *Bueg*, a leg.
> *knospen* stands in South-Tirol for wooden shoes, and *fokazie* for cakes used at Eastertide. (*Focaccia* is used for 'cake' in many parts of Italy, and 'dar pan per focaccia' is equivalent to 'tit for tat' all over the Peninsula.)

It remains only to excuse myself for the spelling of the word Tirol. I have no wish to incur the charge of 'pedantry' which has heretofore been laid on me for so writing it. It seems to me that, in the absence of any glaring mis-derivation, it is most natural to adopt a country's own nomenclature; and in Tirol, or by Tirolean writers, I have never seen the name spelt with a *y*. I have not been able to get nearer its derivation than that the Castle above Meran, which gave it to the whole principality, was called by the Romans, when

[1] P. 79.

they rebuilt it, Teriolis. Why they called it so, or what it was called before, I have not been able to learn. The English use of the definite article in naming Tirol is more difficult to account for than the adoption of the *y*, in which we seem to have been misled by the Germans. We do not say ' *the* France ' or ' *the* Italy ; ' even to accommodate ourselves to the genius of the languages of those countries, therefore, that we should have gone out of our way to say ' *the* Tyrol ' when the genius of that country's language does *not* require us so to call it, can have arisen only from a piece of carelessness which there is no need to repeat.

CONTENTS.

CHAPTER I.

VORARLBERG.

Introductory remarks on the use of myths, legends, and traditions; their imagery beyond imitation; have become a study; now a science; Prof. M. Müller; Rev. G. W. Cox—Karl Blind on attractions for the English in Germanic mythology; mythological persons of Tirol—Mythological symbols in art; in poetry; Dante on popular traditions; their record of thoughts and customs; Tullio Dandolo; Depping; Tirolean peasants . . . PAGE 1

Our introduction to Tirol—Excursions round Feldkirch; the Katzenthurm; St. Fidelis; St. Eusebius—Rankweil—Fridolins-kapelle —Valduna—S. Gerold—Route into Tirol by Lindau—Bregenz, birthplace of Flatz—Legend of Charlemagne; of Ulrich and Wendelgard—Ehreguota—Riedenberg school—the natural preserves of Lustenau—Merboth, Diedo, and Ilga—Embs; its chronicles; Swiss embroidery; Sulphur baths; Jews' synagogue—Lichtenstein; Vaduz; Hot sulphur-baths of Pfäffers; Taminaschlund; Luziensteig 12

From Feldkirch to Innsbruck—The Pass of Frastanz; Shepherd lad's heroism; the traitor's fate—S. Joder and the Devil—Bludenz—Montafon; who gave it its arms—Prazalanz—The Tear-rill; Kirschwasser—Dalaas—Silberthal—Das Bruederhüsle—Engineering of the Arlberg pass—Stanzerthal—Hospice of St. Christof—Wiesburg—Ischgl; its 'skullery'—Landeck—Legend of Schrofenstein—Sharpshooter's monument—Auf dem Fern—Nassereit—Tschirgants Branch road to Füssen—Plansee—Lechthal—Imst—Pitzthal—Growth of a modern legend—Heiterwang—Ehrenberger Klauze Archenthal—Vierzehn Nothhalfer 24

A border adventure; our party; our plans; our route—Aarau—Rorschach; its skeleton-Caryatidæ—Oberriet—Our luggage overpowers the station-master—Our wild colt—Our disaster—Our walk—Our embroideress guide—The Rhine ferry—The Rhætian Alps—Altenstadt—Schattenburg—British missionaries to Tirol—Feldkirch, festa, costumes—Our luggage again—Our new route—Our postilion—The Stase-saddle—The Devil's House—The Voralbergerghost . . . . PAGE 39

CHAPTER II.

NORTH TIROL—UNTERINNTHAL—(RIGHT INN-BANK).

KUFSTEIN TO ROTTENBURG.

Kufstein—Pienzenau's unlucky joke—Ainliffen—Rocsla Sandor; the Hungarian lovers — National anthem — Thierberg — A modern pilgrim—Der Büsser—Public memorials of religion—Zell—Ottokapelle—Kundl—S. Leonhard auf der Wiese; its sculptures—Henry II.'s vow—The Auflänger-Bründl—Rattenberg—Rottenburg—St. Nothburga; her integrity, charity, persecution, patience, piety, observance of Sunday; judgment overtakes Ottilia: Nothburga's restoration; legend of her burial—Henry VI. of Rottenburg and Friedrich mit der leeren Tasche—Character of each—Henry's literary tastes; his mysterious fate—The fire spares Nothburga's cell—Mining legend 53

CHAPTER III.

NORTH TIROL—UNTERINNTHAL—(RIGHT INN-BANK).

THE ZILLERTHAL.

The Zillerthal—Conveyances—Etruscan remnant—Thurnegg and Tratzberg across the river—Strass—Corn or coin?—The two churches of Schlitters—Castles of the Zillerthal—The peace of Kropfsberg—'The only Fügen'—The patriot Riedl—Zell—Expulsion of Lutherans—Hippach—Hainzenberg; ultra co-operative gold mines—Mayrhof — Garnet mills — Mariä-Rastkapelle — Hulda — Tributary valleys—Duxerthal—Hinter-Dux—Hardiness of the people—Legends of the frozen wall—Dog's-throat valley—The Devil's path—The Zemmer glacier—Schwarzensteingrund 79

CHAPTER IV.

NORTH TIROL—UNTERINNTHAL—(RIGHT INN-BANK).

ZILLERTHAL CUSTOMS—THE WILDSCHÖNAU.

Zillerthal customs—Games—Spirits play with gold skittles—Pedlar of Starkenberg—Dances : Schnodahüpfl : Hosennagler—Cow-fights —Kirchtag—Primizen and Sekundizen—Carneval—Christnacht— Kloubabrod—Sternsingen—Gömacht—Weddings—Zutrinken—Customs of other valleys—The cat, patron of courtship . PAGE 92

Kundl again—Wiltschenau—Niederaich — Kundlburg—Oberau—Niederrau—Thierberg—Silver-mines—Legends of dwarfs and Knappen —Moidl and the gold-cave—Legend of the Landmark—Der Umgehende Schuster—Perchtl, Pilate's wife—Comparative mythologists— Wodin, Wilder Jäger, Wilhem Tell—Symbolism in tales of enchanted Princesses—Perahta, the daughter of Dagha—Brixlegg—Burgleckner—Claudia de' Medici—Biener's dying challenge—The Bienerweible—Sandbichler, the Bible-commentator 110

CHAPTER V.

NORTH TIROL—UNTERINNTHAL.

LEFT INN-BANK.

Jenbach—Wiesing—Thiergarten—Kramsach—Brandenberger Ache— Voldepp— The Moosertal — The Mariathal—Rheinthalersee — Achenrain—Mariathal, village and ruined Dominican convent— Georg von Freundsberg—The Brandenbergerthal—Steinberg—Heimaththal, Freiheitthal—The gold-herds of the Reiche Spitze—Die Kalte Pein—Mariastein—The irremovable image—Jenbach—Wiesing—The Thiergarten—The Achenthal—The Käsbachthal—The Blue Achensee—Skolastica—Pertisau—Buchau, Nature's imitation fortress—Tegernsee—The Achen-pass—The judgment of Achensee —Playing at ball in St. Paul's cathedral—Legend of Wildenfeld— Eben—The escape of the vampire—Stans—Joseph Arnold—Tirolean artists—The Stallenthal—St. Georgenberg—Unsere liebe Frau zur Linde—Viecht, Benedictine monastery, library, sculpture—Vomperthal—Sigmundslust—Sigismund the Monied—Terfens—Marialarch—Volandseck—Thierberg—S. Michael's—S. Martin's—The Gnadenwald—Baumkirchen—Fritzens—External tokens of faith—

The holy family at home—Frost phantoms—Hall; Münzthurm; Sandwirthszwanziger; salt-works; Speckbacher; Waldaufischer-Kapelle; S. Saviour's; institutions of Hall—Johanniswürmchen; Bauernkrieg—Excursions round Hall; the Salzberg; the explorations of the 'Fromme Ritter;' grandeur of the salt-mines; salt-works; visit of Hofer and Speckbacher; the Salzthal—Absam; the dragons of Schloss Melans; Count Spaur's ride to Babylon; combat with the toad—Max Müller on legends—The image on the window-pane; the Gnadenmutter von Absam; Stainer the violin-maker—Mils—Grünegg—Schneeberg—The Gnadenwald—The Glockenhof; the Glockengiesser; his temptation, condemnation, and dying request—The Loretokirche—Heiligenkreuz—Taur—Thürl—The Kaisersaüle—St. Romedius, St. Vigilius and the bear; the spectre priest—Rum, landslip . . . . PAGE 125

CHAPTER VI.

NORTH TIROL—UNTERINNTHAL—(RIGHT INN-BANK).

SCHWATZ.

Schwatz, its situation; effigy of S. John Nepomuk; his example; the village frescoes; a hunt for a breakfast; the lessons of traveller's fare; market; church; its size disproportioned to the population; the reason of this—Schwatz a Roman station; silver-mines; prosperity; importance; influence of miners of Saxony; reformation; riots; polemical disputes; decline; copper and iron works; other industries; misfortunes. History of the parish church; peculiar construction; the Knappenhochaltar; monuments; Hans Dreyling; altar-pieces; Michaels-kapelle; its legend; churchyard; its reliquary and holy oil; the Robler and the gossip's corpse; penance and vision of the unmarried—Franciscan church—characteristics of the inns; singular use of the beds; guitar playing—Blessed Sacrament visits the sick —Freundsberg; the ruined castles of Tirol; Georg von Freundsberg; his prowess, strength, success; devotion of his men; sung of as a hero; his part in the siege of Rome, sudden death, and ruin of his house; tower; chapel—Weird-woman; her story; her legends; Oswald Milser of Seefeld; the bird-catcher of the Goaslahn; strange birds; chamois; the curse of the swallow—Hospital; chapel—Tobacco; factory girls at benediction—Pews in German churches . . 168

CHAPTER VII.

NORTH TIROL—UNTERINNTHAL—(RIGHT INN-BANK).

EXCURSIONS FROM SCHWATZ.

Falkenstein; exhausted mines; religious observances of miners; tokens of their craft— Buch — Margareth — Galzein — Kugelmoos—The Schwaderalpe—The Kellerspitze—Troi—Arzberg—Heiligenkreuz-kapelle—Baierische-Rumpel—Pill—The Woerthal, Schloss Rottenberg; its spectre warder—The Kolsassthal—Wattens—Walchen—Mols—The Navisthal—Lizumthal; the Blue Lake—Volders—Voldererthal—Hanzenheim—Friedberg—Aschbach, why it is in the parish of Mils—Hippolitus Guarinoni, page to St. Charles, physician of the poor; religious zeal; church of St. Charles, Servitenkloster, the Stein des Gehorsams; analogous legend—Rinn; S. Anderle's martyrdom; the Judenstein; lettered lilies—Aversion to Jews—Voldererbad—Ampass—Lans—The Patscherkofl—The Lansersee; the poor proprietor and the unjust noble—Sistrans; legend of its champion wrestler—Heiligenwasser . . . . PAGE 200

CHAPTER VIII.

NORTH TIROL—THE INNTHAL.

INNSBRUCK.

Our greeting; characteristics of the people; Innsbruck's treatment of Kaiser Max; the Œstereichisher hof; our apartment; mountain view; character of the town; its history—Wilten; the minster; myth of Haymon the giant; his burial-place; parish church; Marienbild unter den vier Säulen; relic of the thundering legion—First record of Innsbruck; chosen for seat of government; for residence by Friedl mit der leeren Tasche—Character of Tirolean rulers—the Goldene-Dachl-Gebäude—Sigismund the Monied; his reception of Christian I.; condition of Tirol in his time; his castles; abdication—Maximilian; builds the Burg; magnificence of his reign; legends of him; his decline—Charles Quint; cedes Tirol to Ferdinand I.; his wise administration; quiets popular agitation; Charles Quint's visits to Innsbruck; attacked by Maurice, Elector of Saxony; carried into Carinthia in a litter; death of Maurice—Ferdinand I., the Hof-Kirche; Maximilian's cenotaph; its bas-relief; statues; Mirakel-Bild des H. Anton; Fürstenchor; abjuration of Queen Christina—Introduction of Jesuits; results—The 'Fromme Siechin'—Ferdinand II.; his peaceful tastes; romantic attachment; Philippine Welser; ménage at Schloss Ambras; collections; curiosities; portraits; Philippine's end . . . . PAGE 225

CHAPTER IX.

NORTH TIROL—THE INNTHAL.

INNSBRUCK (continued).

Wallenstein's vow—Theophrastus Paracelsus; his mysterious dealings—The Tummelplatz—The Silberne Kapelle—Earthquake and dearth; their lessons—Ferdinand's devotion to the Blessed Sacrament; analogous legend of Rudolf of Hapsburg—Ferdinand's second marriage—The Capuchin Church—Maximilian the Deutschmeister; introduces the Servites—Paul Lederer—Maximilian's hermitage—S. Lorenzo of Brindisi—Dreiheiligkeitskirche—Provisions against ravages of the Thirty Years' War—The Siechenhaus—Leopold V.; dispensed from his episcopal jurisdiction and vows; Marries Claudia de' Medici—Friedrich v. Tiefenbach—Festivities at Innsbruck—The Hofgarten—Kranach's Madonna, Mariähülfskirche built to receive it; translation to the Pfarr-kirche under Ferdinand Karl—Ferdinand Karl—Regency of Claudia de' Medici; administrative ability; Italian influences—Sigismund Franz—Claudia Felicita—Charles of Lotharingia—War of succession; Bavarian inroad of 1703; the Pontlatzerbrücke; Baierische-Rumpel—St. Annensäule—Joseph I.—Karl Philipp; builds the Land-haus and gymnasium, restores the Pfarrkirche; stucco and marble decorations; frescoes; preservation of Damian Asam—Strafarbeitshaus—Church of S. John Nepomuk; his popularity; canonisation—Maria Theresa; her partiality for Innsbruck; example; Prussian prisoners; marriage of Leopold; death of Francis I.; the Triumphpforte, the Damenstift—Joseph II.—Archduchess Maria Elizabeth—Pius VI. passes through Innsbruck—Leopold II.—Repeal of Josephinischen measures—Francis II.—Outbreak of the French revolution—Das Mädchen v. Spinges—The Auferstehungsfeier—Archduchess Maria Elizabeth—Gottesacker—Treaty of Pressburg—'The Year Nine'—Andreas Hofer—Peace of Schönbrunn—Speckbacher; successes at Berg Isel; Hofer as Schutzen-Kommandant; his moderation, simplicity, subordination; his betrayal; last hours;

firmness; execution—Restoration of Austrian rule—Hofer's monument—Tirolese loyalty in 1848—The Ferdinandeum; its curiosities—Early editions of German authors—Paintings on cobweb—The Schiess-stand—Policy of the Viennese Government, constitutional opposition of Tirol—Population of Innsbruck . . PAGE 265

CHAPTER X.

NORTH TIROL—OBERINNTHAL.

INNSBRUCK TO ZIRL AND SCHARNITZ—INNSBRUCK TO THE LISENS-FERNER.

Excursions from Innsbruck—Mühlau; new church; Baronin Sternbach—Judgment of Frau Hütt—Büchsenhausen—Weierburg—Mariä-Brunn—Hottingen; monuments in the Friedhof—Schloss Lichtenthurm—The Höttingerbild; the student's Madonna; stalactites—Excursion to Zirl—Grossen Herr-Gott Strasse—Kranebitten—The Schwefelloch — The Hundskapelle — The Zirlerchristen — Grosssolstein—The Martinswand; danger of the Emperor Maximilian; Collin's ballad; who led the Kaiser astray?—His importance in Europe; efforts to rescue him; the Blessed Sacrament visits him; unknown deliverer—Martinsbühl—Traditions of Kaiser Max—Zirl—Fragenstein; its hidden treasure—Leiten—Reit—Seefeld—The Heilige Blutskapelle—The Seekapelle—Scharnitz—Isarthal—Porta Klaudia—Dirstenöhl—The beggar-woman's prayer; vision of the peasant of Dorf 310

Unter-Perfuss—Selrainthal—The Melach—Rothenbrunn—Fatscherthal—The Hohe Villerspitz—Sonnenberg—Magdalenen-Bründl—Character of the Selrainthalers—Ober-Perfuss; Peter Anich—Kematen—Völs; the Blasienberg; S. Jodok—The Galwiese—The Schwarze-Kreuzkapelle; Hölzl's vow—Ferneck—Berg Isel—Noise of the rifle practice—Count v. Stachel—Natters and Mutters—Waidburg—The Nockspitze—Götzens—Schloss Völlenberg; Oswald v. Wolkenstein—Birgitz—Axams—The Sendersthal 329

CHAPTER XI.

WÄLSCH-TIROL.

THE WÄLSCHEROLISCHE-ETSCHTHAL AND ITS TRIBUTARY VALLEYS.

Val di Lagarina—Borghetto—Ala—Roveredo—Surrounding castles—Dante at Lizzana—The Slavini di S. Marco—La Busa del Barbaz;

its myths—Serravalle—Schloss Junk—The Madonna del Monte—
Industries—Chapel of S. Columban—Trent, Festa of St. Vigilius;
comparison between Trent and Rome; the Domkirche; its notabilia;
Sta. Maria Maggiore; seat of the council; assenting crucifix; cen-
tenary celebration; legend of the organ-builder—Church of St.
Peter; Chapel of S. Simonin; club; museum; Paluzzi; Palazzo
Zambelli, Teufelspalast; its legend; General Gallas—The Madonna
alle Laste; view of Trent—Dos Trento—St. Ingenuin's garden; St.
Albuin's apples—Lavis—French spoliation—Restitution—Wälsch
Michel   .   .   .   .   PAGE 340

Tributary valleys—Val di Non; Annaunia—Rochetta Pass Wälschmetz
—Visiaun—Spaur Maggiore—Denno—Schloss Belasis—The Seidenbaum—Tobel Wild-see—Cles; Tavola Clesiana; Roman remains;
the Schwarzen Felder—SS. Sisinus, Martyrius and Alexander—
Val di Sole—Livo—Magras; Val di Rabbi; San Bernardo—Malè—
Charles Quint's visit — Pellizano—Val di Pejo—Cogolo—Corno
de' tre Signori—Val Vermiglio—Tonale; the witches' sabbath
there—Tregiovo—Cloz—U-Liebe Frau auf dem Gampen—Fondo—
Sanzeno—Legend of the three brothers; mithraic bas-relief—The
Tirolean Petrarch—St. Romediusthal; legend of St. Romedius;
angelic consecration; conversion of the false penitents; extraordinary construction and arrangement of the building; romantic
situation; fifteen centuries of uninterrupted veneration—Castel
Thun; attachment of the people to the family; a Nonesade;
aqueduct—Dombel; its Etruscan key; its import   .   .   358

The Avisiothal—Val di Cembra; its inaccessibility—Altrei; presentation of colours—Fleimserthal; Cavalese; its church a museum of
Tirolese Art; local parliament; legend of its site; handsome new
church—Fassathal—Moena—Analogous English and French traditions—Marriage customs of the valley—The Feuriger Verräther—
Vigo—The Marmolata; its legends—St. Ulrich   .   .   .   374

CHAPTER XII.

WÄLSCH-TIROL.

VAL SUGANA—GIUDICARIA—FOLKLORE.

Val Sugana—Baselga—The Madonna di Pinè; legend of the Madonna
di Caravaggio—Pergine; miners; the Canoppa—The Schloss—
Marriage customs of the valley—Lake Caldonazzo—St. Hermes at
Calzeranica—Bosentino—Nossa signora del Feles—The sleeper of

Vallo del Orco—Caldonazzo—Lafraun ; legend of the disunited brothers—Borgo, the Italian Meran—Franciscan convent ; Castel Telvana ; dangers of a carneval procession ; Count Welsburg's vow—Gallant border defences—Stalactite caves of Costalta—Sette Comuni—Castelalto—Strigno — Castelrotto — Cima d'Asta—Quarazza garnet quarry—Ivano—Grigno ; Legend of St. Udalric—Castel Tesino—Canal San Bovo to Primiero—Tale of Virginia Loss ; humble heroism—Le Tezze ; modern heroes . . . PAGE 382

Judicarien ; its divisions—Castel Madruzz ; Cardinal Karl Madruzz ; his dispensation ; its conditions—Abraham's Garden—Sta. Massenza ; Bishop's Summer Palace—Loreto-kapelle—The Rendenathal ; St. Vigilius ; his zeal ; early admission to the episcopate ; missionary labours : builds churches ; overthrows idols ; his stoning ; his burial ; the rock cloven for his body to pass ; the Acqua della Vela ; the bread of Mortaso—S. Zulian ; his legend ; his penitence—Caresolo ; its frescoes ; another memorial of Charles Quint ; his estimation of Jews—New churches—Legends of Condino and Campiglio—Riva on the Garda-see ; its churches ; its olive branches—The Altissimo di Nago ; view from S. Giacomo ; optical illusion—Brentonico—The Ponte delle Streghe—Mori ; tobacco cultivation . . . 400

Character of Wälsch-Tirol folklore—Orco-Sagen ; his transformations in many lands ; transliterations of his name in Tirol—The Salvan and Gannes ; perhaps Etruscan genii—Salvanel ; Bedelmon ; Salvadegh—The Beatrik, identified with Dietrich von Bern—The Angane—What came of marrying an Angana—The focarelli of Lunigiana—The Filò—Froberte—Donna Berta dal nas longh—The discriminating Salvan—The Angana's ring ; tales of the Three Wishes and the Faithful Beasts ; legend of the Drei Feyen of Thal Vent—Legend of St. Kummerniss ; her effigy in Cadore ; the prevailing minstrel—Turlulù—Remnants of Etruscan language—Storielle da rider'—The bear-hunters—The horrible snail— How to make a church tower grow—Social customs perhaps derived from Etruscan ; similar to those of Lombardy and Lunigiana—All Souls' Day ; feast of Sta. Lucia ; Christmas ; St. Anthony's Day ; Carneval ; Giovedi de' Gnocchi ; St. Urban—Popular sayings about thunder, crickets, brambles, cockchafers, swallows, scorpions—Astronomical riddles 408

LIST OF ILLUSTRATIONS.

Kufstein . . *Frontispiece.*

MAPS.

The Valleys of Tirol   *to face p.* 12

Unterinnthal and Neighbourhood of Innsbruck   ,,   53

Wälsch-Tirol . . .   ,,   341

THE
VALLEYS OF TIROL

THEIR TRADITIONS AND CUSTOMS.

CHAPTER I.

VORARLBERG.

> . . . . . *Everywhere*
> *Fable and Truth have shed, in rivalry,*
> *Each her peculiar influence. Fable came,*
> *And laughed and sang, arraying Truth in flowers,*
> *Like a young child her grandam. Fable came,*
> *Earth, sea, and sky reflecting, as she flew,*
> *A thousand, thousand colours not their own.*—ROGERS.

'TRADITIONS, myths, legends! what is the use of recording and propagating the follies and superstitions of a bygone period, which it is the boast of our modern enlightenment to have cast to the winds?'

Such is the hasty exclamation which allusion to these fantastic matters very frequently elicits. With many they find no favour because they seem to yield no profit; nay, rather to set up a hindrance in the way of progress and culture.

Yet, on the other hand, in spite of their seeming foolishness, they have worked themselves into favour with very various classes of readers and students. There is an audacity in their imagery which no mere sensation-writer could attempt without falling Phaeton-like from his height; and they plunge us so hardily into a world of their own, so preposterous and so unlike ours, while all the time describing it in a language we can understand without effort, that no one who seeks occasional relief from modern monotony but must experience refreshment in the weird excursions their jaunty will-o'-the-wisp dance leads him. But more than this; their sportive fancy has not only charmed the dilettante; they have revealed that they hold inherent in them mysteries which have extorted the study of deep and able thinkers, one of whom [1] insisted, now some years ago, that ' by this time the study of popular tales has become a recognized branch of the studies of mankind;' while important and erudite treatises from his own pen and that of others[2] have elevated it further from a study to a science.

All who love poetry and art, as well as all who are interested in the study of languages or races, all who have any care concerning the stirrings of the human mind in its search after the supernatural and the infinite, must confess to standing largely in debt, in the absence of more positive records of the earliest phases of thought, to these various mythologies.

[1] Professor Max Müller, *Chips from a German Workshop*.
[2] Rev. G. W. Cox, Prof. De Gubernatis, Dr. Dasent, &c.

Karl Blind, in a recent paper on 'German Mythology,'[1] draws attention to some interesting considerations why the Germanic traditions, which we chiefly meet with in Tirol, should have a fascination for us in this country, in the points of contact they present with our language and customs. Not content with reckoning that 'in the words of the Rev. Isaac Taylor we have obtruded on our notice the names of the deities who were worshipped by the Germanic races' on every Tuesday, Wednesday, Thursday, and Friday of our lives, as we all know, he would even find the origin of 'Saturday' in the name of 'a god "Sætere" hidden, (a malicious deity whose name is but an *alias* for Loki,) of whom, it is recorded, that once at a great banquet he so insulted all the heavenly rulers that they chained him, Prometheus-like, to a rock, and made a serpent trickle down its venom upon his face. His faithful wife Sigyn held a cup over him to prevent the venom reaching his face, but whenever she turned away to empty the cup his convulsive pains were such that the earth shook and trembled. . . . Few people now-a-days, when pronouncing the simple word "Saturday," think or know of this weird and pathetic myth.[2] . . . When we go to Athens we easily think of the Greek goddess Athene, when we go to Rome we are reminded of Romulus its mythic founder. But

---

[1] In the *Contemporary Review* for March 1874.

[2] Mr. Cox had pointed it out before him, however, and more fully, *Mythology of the Aryan Nations*, ii. 200.

when we go to Dewerstone in Devonshire, to Dewsbury in Yorkshire, to Tewesley in Surrey, to Great Tew in Oxfordshire, to Tewen in Herefordshire—have a great many of us even an inkling that these are places once sacred to Tiu, the Saxon Mars? When we got to Wednesbury, to Wanborough, to Woodnesborough, to Wembury, to Wanstrow, to Wanslike, to Woden Hill, we visit localities where the Great Spirit Wodan was once worshipped. So also we meet with the name of the God of Thunder in Thudersfield, Thundersleigh, Thursleigh, Thurscross, Thursby, and Thurso. The German Venus Freia is traceable in Fridaythorpe and Frathorpe, in Fraisthorpe and Freasley. Her son was Baldur, also called Phol or Pol, the sweet god of peace and light; his name comes out at Balderby, Balderton, Polbrook, Polstead and Polsden. Sætere is probably hidden in Satterleigh and Satterthwaite; Ostara or Eostre, the Easter goddess of Spring, appears in two Essex parishes, Good Easter and High Easter, in Easterford, Easterlake and Eastermear. Again Hel, the gloomy mistress of the underworld, has given her name to Hellifield, Hellathyrne, Helwith, Healeys and Helagh—all places in Yorkshire, where people seem to have had a particular fancy for that dark and grimy deity. Then we have Asgardby and Aysgarth, places reminding us of Asgard, the celestial garden or castle of the Æsir—the Germanic Olympus. And these instances might be multiplied by the hundred, so full is England to this day of the vestiges of

Germanic mythology. Far more important is the fact that in this country, just as in Germany, we find current folk-lore; and quaint customs and superstitious beliefs affecting the daily life, which are remnants of the ancient creed. A rime apparently so bereft of sense as

> Ladybird, ladybird, fly away home!
> Thy house is on fire!
> Thy children at home!

can be proved to refer to a belief of our forefathers in the coming downfall of the universe by a great conflagration. The ladybird has its name from having been sacred to our Lady Freia. The words addressed to the insect were once an incantation—an appeal to the goddess for the protection of the soul of the unborn, over whom in her heavenly abode she was supposed to keep watch and ward, and whom she is asked to shield from the fire that consumes the world. . . . . If we ever wean men from the crude notions that haunt them, and yet promote the enjoyment of fancies which serve as embellishing garlands for the rude realities of life, we cannot do better than promote a fuller scientific knowledge of that circle of ideas in which those moved who moulded our very speech. We feel delight in the conceptions of the Greek Olympus. Painters and poets still go back to that old fountain of fancy. Why should we not seek for similar delight in studying the figures of the Germanic Pantheon, and the rich folk-lore connected with them? Why should that powerful Bible of the Norse religion, which contains such a wealth of striking ideas and descriptions in language the most picturesque, not be as much perused as the Iliad, the Odyssey, or the Æneid? Is it too much to say that many even of those who know of the Koran, of the precepts of Kou-fu-tsi and of Buddha, of the Zendavesta and the Vedas, have but the dimmest notion of that grand Germanic Scripture? . . .

'Can it be said that there is a lack of poetical conception in the figure of Wodan or Odin, the hoary ruler of the winds and the clouds, who, clad in a flowing mantle, careers through the sky on a milk-white horse, from whose nostrils fire issues, and who is followed at night by a retinue of heroic warriors whom he leads into the golden shield-adorned Walhalla? Is there a want of artistic delineation in Freia—an Aphrodite and Venus combined, who changes darkness into light wherever she appears—the goddess with the streaming golden locks and siren voice, who hovers in her sun-white robe between heaven and earth, making flowers sprout along her path and planting irresistible longings in the hearts of men? Do we not see in bold and well-marked outline the figure of the red-bearded, steel-handed Thor, who rolls along the sky in his goat-drawn car, and who smites the mountain giants with his magic hammer? Are these mere spectres without distinct contour? . . . are they not, even in their uncouth passions, the representatives of a primitive race, in which the pulse throbs with youthful freshness? Or need I allude to that fantastic theory of minor deities, of fairies and wood-women, and elfin and pixies and cobolds, that have been evolved out of all the forces of Nature by the Teutonic mind, and before whose bustling crowd even Hellenic imagination pales?

'Then what a dramatic power has the Germanic mythology! The gods of classic antiquity have been compared to so many statues ranged along a stately edifice . . . in the Germanic view all is active struggle, dramatic contest, with a deep dark background of inevitable fate that controls alike gods and men.'

Such are the Beings whom we meet wandering all over Tirol; transformed often into new personalities, invested with new attributes and supplemented with many a mysterious companion, the offspring of an imagination informed by another order of thought, but all of them more living, and more readily to be met with, than in any part of wonder-loving Germany itself.

Apart from their mythological value, how large is the debt we owe to legends and traditions in building up our very civilization. Their influence on art is apparent, from the earliest sculptured stones unearthed in India or Etruria to the latest breathing of symbolism in the very reproductions of our own day. In poetry, no less a master than Dante lamented that their influence was waning at the very period ascribed a few years ago as the date of their taking rise. Extolling the simpler pursuits and pleasures of his people at a more primitive date than his own, 'One by the crib kept watch,' he says, 'studious to still the infant plaint with words which erst the parents' minds diverted; another, the flaxen maze upon the distaff twirling, recounted to her household, tales of Troy, Fiesole, and Rome.'[1] Their work is patent in his own undying pages, and in those of all true poets before and since.

Besides all this, have they not preserved to us, as in a registering mirror, the manners and habits of thought of the ages preceding ours? Have they not served to record as well as to mould the noblest aspirations of those who have gone before? 'What are they,' asks an elegant Italian writer of the present day,[2] treating, however, only of the traditions of the earliest epoch of Christianity, ' but narratives woven beside the chimney, under the tent, during the halt of the caravan, embodying as in a lively picture the popular customs of the apostolic ages, the interior life of the rising (*nascente*) Christian society? In them we have a delightful opportunity of seeing stereotyped the great transformation and the rich source of ideas and sentiments which the new belief opened up, to illuminate the common people in their huts no less than the patricians in their palaces. Those even who do not please

---

[1] L' una vegghiava a studio della culla,
   E consolando usava l' idioma,
   Che pria li padri e le madri trastulla :
  L' altra traendo alla rocca la chioma
   Favoleggiava con la sua famiglia
   De' Troiani, e di Fiesole, e di Roma.
            DANTE. *Paradiso*, xv. 120 5.

[2] Tullio Dandolo.

to believe the facts they expose are afforded a genuine view of the habits of life, the manner of speaking and behaving—all that expresses and paints the erudition of those men and of those times. Thus, it may be affirmed, they comment beautifully on the Gospels, and in the midst of fables is grafted a great abundance of truth.

'If we would investigate the cause of their multiplication, and of the favour with which they were received from the earliest times, we shall find it to consist chiefly in the need and love of the marvellous which governed the new society, notwithstanding the severity of its dogmas. Neophytes snatched from the superstitions of paganism would not have been able all at once to suppress every inclination for poetical fables. They needed another food according to their fancy. And indeed were they not great marvels (though of another order from those to which they were accustomed) which were narrated to them? The aggregate mass was, however, increased by the way in which they lived and the scarcity of communication; every uncertain rumour was thus readily dressed up in the form of a wonderful fact.

'Again, dogmatic and historical teaching continued long to be oral; so that when an apostle, or the apostle of an apostle, arrived in any city and chained the interest of the faithful with a narration of the acts of Jesus he had himself witnessed or received from the personal narrative of witnesses, his words ran along from mouth to mouth, and each repeater added something, suggested by his faith or by his heart. In this way his teaching constituted itself into a legend, which in the end was no longer the narrative of one, but the expression of the faith of all.

'Thus whoever looks at legends only as isolated productions of a period most worthy of study, without attending to the influence they exercised on later epochs, must even so hold them in account as literary monuments of great moment.'

Nor is this the case only with the earliest legends. The popular mind in all ages has evinced a necessity for filling up all blanks in the histories of its heroes. The probable, and even the merely possible, is idealized; what *might have been* is reckoned *to have* happened; the logical deductions as to what a favourite saint or cobbold *ought to* have done, according to certain fixed principles of action previously ascribed to his nature, are taken to be the very acts he did perform; and thus, even those traditions which are the most transparently human in their origin, have served to show reflected in action the virtues and perfections which it is the boast of religion to inculcate.

A Flemish writer on Spanish traditions similarly remarks, 'Peoples who are cut off from the rest of the world by such boundaries as seas, mountains, or wastes, by reason of the difficulty of communication thus occasioned, are driven to concentrate their attention to local events; and in their many idle hours they work up their myths and tales into poems, which stand them in stead of books, and, in fact, constitute a literature.'[1]

Europe possesses in Tirol one little country at least in whose mountain fastnesses a store of these treasures not only lies enshrined, but where we may yet see it in request. Primitive and unsophisticated tillers of the soil, accustomed to watch as a yearly miracle the welling up of its fruits, and to depend for their hopes of subsistence on the sun and rain in the hand of their Creator, its children have not yet acquired the independence of thought and the habit of referring all events to natural causes, which is generated by those industries of production to which the human agent appears to be all in all. Among them we have the opportunity of seeing these expositions of the supernatural, at home as it were in their contemporary life, supplying a representation of what has gone before, only to be compared to the revelations of deep-cut strata to the geologist, and the unearthing of buried cities to the student of history. It is further satisfactory to find that, in spite of our repugnance to superstition, this unreasoning realization of the supernatural has in no way deteriorated the people. Their public virtues, seen in their indomitable devotion to their country, have been conspicuous in all ages, no less than their heroic labours in grappling with the obstacles of soil and climate; while all who have visited them concur in bearing testimony to their possession of sterling homely qualities, frugality,

[1] Depping, *Romancero, Preface.*

morality, hospitality; and, for that which is of most importance to the tourist, all who have been among them will bear witness to the justice of the remark in the latest Guide-book, that, except just in the more cultivated centres of Innsbruck, Brixen, and Botzen, you need take no thought among the Tiroleans concerning the calls on your purse.

My first acquaintance with Tirol was made at Feldkirch, where I had to pay somewhat dearly for my love of the legendary and the primitive. Our plan for the autumn was to join a party of friends from Italy at Innsbruck, spend some months of long-promised enjoyment in exploring Tirol, and return together to winter in Rome. The arrangements of the journey had been left to me; and as I delight in getting beyond railways and travelling in a conveyance whose pace and hours are more under one's own control, I traced our road through France to Bâle, and then by way of Zurich and Rorschach and Oberriet to Feldkirch (which I knew to be a post-station) as a base of operations, for leisurely threading our mountain way through Bludenz and Landeck and the intervening valleys to Innsbruck.

How our plan was thwarted [1] I will relate presently. I still recommend this line of route to others less encumbered with luggage, as leading through out-of-the-

---

[1] The usual fate of relying on Road-books. Ours, I forget whether Amthor's or Trautwein's, said there was regular communication between Oberriet and Feldkirch, and nothing could be further from the fact, as will be seen a few pages later.

way and unfrequented places. The projected railway between Feldkirch and Innsbruck is now completed as far as Bludenz; and Feldkirch is reached direct by the new junction with the Rorschach-Chur railway at Buchs-station.[1]

Feldkirch affords excursions, accessible for all, to the Margarethenkapf and the St. Veitskapf, from either of which a glorious view is to be enjoyed. The latter commands the stern gorges through which the Ill makes its final struggles before losing its identity in the Rhine —struggles which are often terrific and devastating, for every few years it carries down a whole torrent of pebbles for many days together. The former overlooks the more smiling tracts we traversed in our forced march, locally called the Ardetzen, hemmed in by noble mountain peaks. Then its fortifications, intended at one time to make it a strong border town against Switzerland, have left some few picturesque remains, and in particular the so-called Katzenthurm, named from certain clumsy weapons styled 'cat's head guns,' which once defended it, and which were ultimately melted down to make a chime of peaceful bells. And then it has two or three churches to which peculiar legends attach. Not the least curious of these is that of St. Fidelis, a local saint, whose cultus sprang up as late as the year 1622, when he was laid in wait for and assassinated by certain fanatical reprobates, whose consciences his earnest preaching had disturbed. He was declared a

---

[1] If Pfäffers is visited by rail (see p. 23), it is convenient to take it before Feldkirch.

martyr, and canonized at Rome in 1746. The sword with which he was put to death, the bier on which his body was carried back into the town, and other things belonging to him, are venerated as relics. About eight miles outside the town another saint is venerated with a precisely similar history, but dating from the year 844. This is St. Eusebius, one of a band of Scotch missionaries, who founded a monastery there called Victorsberg, the oldest foundation in all Vorarlberg. St. Eusebius, returning from a pilgrimage one day, lay down to sleep in this neighbourhood, being overtaken by the darkness of night. Heathen peasants, who had resisted his attempts at converting them, going out early in the morning to mow, found him lying on the ground, and one of them cut off his head with his scythe. To their astonishment the decapitated body rose to its feet, and, taking up the head in its hands, walked straight to the door of the monastery, where the brethren took it in and laid it to rest in the churchyard. A little further (reached most conveniently by a by-path off the road near Altenstadt, mentioned below,) is Rankweil. In the church on Our Lady's Mount (Frauenberg) is a little chapel on the north side, where a reddish stone is preserved (*Der rothe Stein in der Fridolinskapelle*), of which the following story is told. St. Fridolin was a Scotch missionary in the seventh century, and among other religious houses had founded one at Müsigen. Two noblemen of this neighbourhood (brothers) held him in great respect, and before dying, one of them, Ursus by name, endowed the convent with all his worldly goods. Sandolf, the other, who did not carry his admiration of the saint to so great a length as to renounce his brother's rich inheritance, disputed the possession, and it was decided that Fridolin must give it up unless he could produce the testimony of the donor. Fridolin went in faith to Glarus, where Ursus had been buried two years before. At his call the dead man rose to his feet, and pushing the grave-stone aside, walked, hand-in-hand, with his friend back to Rankweil, where he not only substantiated Fridolin's statements, but so effectually frightened his brother that he immediately added to the gift all his own possessions also. But the story says that when the judgment requiring him to produce the testimony of the dead was first given, Fridolin went to pray in the chapel of Rankweil, and there a shining being appeared to him, and told him to go to Glarus and call Ursus; and as he spoke Fridolin's knees sank into the 'red stone,' making the marks now seen.[1]

The reason given why this hill is called Our Lady's Mound is, that on it once stood a fortress called Schönberg. Schönberg having been burnt down, its owner, the knight of Hörnlingen, set about rebuilding it; but whatever work his workmen did in the day-time, was destroyed by invisible hands during the night. A pious old workman, too, used to hear a mysterious voice saying

[1] See further quaint details and historical particulars in Vonbun, *Sagen Vorarlbergs*, p. 103-5.

that instead of a fortress they should build a sanctuary in honour of the mother of God. The knight yielded to the commands of the voice, and the church was built out of the ruins of his castle. In this church, too, is preserved a singular antique cross, studded with coloured glass gems, which the people venerate because it was brought down to them by the mountain stream. It is obviously of very ancient workmanship, and an inscription records that it was *repaired* in 1347.

Winding round the mountain path which from Rankweil runs behind Feldkirch to Satteins, the convent of Valduna is reached; and the origin of this sanctuary is ascribed to a legend, of which counterparts crop up in various places, of a hermit who passed half a life within a hollow tree,[1] and acquired the lasting veneration of the neighbouring people.

Another mountain sanctuary which received its veneration from the memory of a tree-hermit, is S. Gerold, situated on a little elevation below the Hoch Gerach, about seven miles on the east side of Feldkirch. It dates from the tenth century. Count Otho, Lord of Sax in the Rhinethal, was out hunting, when the bear to which he was giving chase sought refuge at the foot of an old oak tree, whither his dogs durst not follow it. Living as a hermit within this oak tree Count Otho found his long lost father, S. Gerold, who years before had forsaken his throne and found there a life of contemplation in the wild.[2] The tomb of the saint and

[1] Vonbun, pp. 113-4.
[2] Historical particulars in Vonbun, pp. 110-1.

his two sons is to be seen in the church, and some curious frescoes with the story of his adventures.

Another way to be recommended for entering Vorarlberg is by crossing Lake Constance from Rorschach to Lindau, a very pleasant trajet of about two hours in the tolerably well-appointed, but not very swift lake-steamers. Lindau itself is a charming old place, formed out of three islands on the edge of the lake; but as it is outside the border of Tirol, I will only note in favour of the honesty of its inhabitants, that I saw a tree laden with remarkably fine ripe pears overhanging a wall in the principal street, and no street-boy raised a hand to them.

The first town in Tirol by this route is Bregenz, which reckons as the capital of Vorarlberg. It may be reached by boat in less than half an hour. It is well situated at the foot of the Gebhartsberg, which affords a most delightful, and in Tirol widely celebrated, view over Lake Constance and the Appenzel mountains and the rapid Rhine between; and here, at either the Post Hotel or the Black Eagle, there is no lack of carriages for reaching Feldkirch. Bregenz deserves to be remembered as the birth-place of one of the best modern painters of the Munich-Roman school, Flatz, who I believe, spends much of his time there.

Among the objects of interest in Bregenz are the Capuchin Convent, situated on a wooded peak of the Gebhardsberg, founded in 1636; on another peak, S. Gebhard auf dem Pfannenberge, called after a bishop of Constance, who preached the Christian faith in the neighbourhood, and was martyred. Bregenz has an ancient history and high lineage. Its lords, who were powerful throughout the Middle Ages, were of sufficiently high estate at the time of Charlemagne that he should take Hildegard, the daughter of one of them, to be his wife, and there is a highly poetical popular tale about her. Taland (a favourite name in Vorarlberg) was a suitor who had, with jealous eye, seen her given to the powerful Emperor, and in the bitterness of his rejected affection, so calumniated her to Charlemagne, that he repudiated her and married Desiderata, the Lombard princess.[1] Hildegard accepted her trial with angelic resignation, and devoted her life to tending pilgrims at Rome. Meantime Taland, stricken with blindness, came to Rome in penitential pilgrimage, where he fell under the charitable care of Hildegard. Hildegard's saintly handling restored his sight—not only that of his bodily eyes, but also his moral perception of truth and falsehood. In reparation for the evil he had done, he now led her back to Charlemagne, confessed all, and she was once more restored to favour and honour. Bregenz has also another analogous and equally beautiful legend. One of its later counts, Ulrich V., was supposed by his people to have died in war in Hungary, about the year 916. Wendelgard, his wife, devoted her widowhood to the cloistral life, but took the veil under the condition that she should every

[1] Vonbun, pp. 86–7.

year hold a popular festival and distribution of alms in memory of her husband. On the fourth anniversary, as she was distributing her bounty, a pilgrim came forward who allowed himself the liberty of kissing the hand which bestowed the dole. Wendelgard's indignation was changed into delight when she recognized that the audaciously gallant pilgrim was no other than her own lord, who, having succeeded in delivering himself from captivity, had elected to make himself thus known to her. Salomo, Bishop of Constance, dispensed her from her vow, and Ulrich passed the remainder of his life at Bregenz by her side. Another celebrated worthy of Bregenz, whose name must not be passed over, is 'Ehreguota' or 'Ehre Guta,' a name still dear to every peasant of Vorarlberg, and which has perpetuated itself in the appellation of Hergotha, a favourite Christian name there to the present day. She was a poor beggar-woman really named Guta, whose sagacity and courage delivered her country people from an attack of the Appenzell folk, to which they had nearly succumbed in the year 1408; it was the 'honour' paid her by her patriotic friends that added the byname of 'Ehre,' and made them erect a monument to her. One of the variants of the story makes her, instead of a beggar-woman, the beautiful young bride of Count Wilhelm of Montfort-Bregenz; some have further sought to identify her with the goddess Epona.

Pursuing the journey southwards towards Feldkirch, every step is full of natural beauty and legendary interest. At first leaving Bregenz you have to part company with Lake Constance, and leave in the right hand distance the ruins of Castle Fussach. On the left is Riedenberg, which, if not great architecturally, is interesting as a highly useful institution, under the fostering care of the present Empress of Austria, for the education of girls belonging to families of a superior class with restricted means. From Fussach the road runs parallel to the Rhine; there is a shorter road by Dornbirn, but less interesting, which joins it again at Götzis, near Hohenembs. The two roads separate before Fussach at Wolfurth, where there is an interesting chapel, the bourne of a pilgrimage worth making if only for the view over the lake. The country between S. John Höchst and Lustenau is much frequented in autumn for the sake of the shooting afforded by the wild birds which haunt its secluded recesses on the banks of the Rhine at that season. At Lustenau there is a ferry over the Rhine.

The favourite saints of this part of the country are Merboth, Diédo, and Ilga—two brothers and a sister of a noble family, hermit-apostles and martyrs of the eleventh century. Ilga established her hermit-cell in the Schwarzenberg, just over Dornbirn, where not only all dainty food, but even water, was wanting. The people of Dornbirn also wanted water; and though she had not asked the boon for herself, she asked it for her people, and obtained from the hard rock, a miraculous spring of sparkling water which even the winter cold could not freeze. Ilga used to fetch this water for her own use, and carry it up the mountain paths *in her apron*. One day she spilt some of it on the rock near her cell on her arrival, and see! as it touched the rock, the rock responded to the appeal, and from out there flowed a corresponding stream, which has never ceased to flow to this day.

The most important and interesting spot between Bregenz and Feldkirch, is Embs or Hohenembs, with its grand situation, its picturesque buildings and its two ruined castles, which though distinguished as Alt and Neu Hohenembs, do not display at first sight any very great disparity of age; both repay a visit, but the view from Alt Hohenembs is the finer. The virtues and bravery of the lords of Hohenembs have been duly chronicled. James Von Embs served by the side of the chevalier Bayard in the battle of Ravenna, and having at the first onset received his death wound, raised himself up again to pour out his last breath in crying to his men, 'The King of France has been our fair ally, let us serve him bravely this day!' His grandson, who was curiously enough christened James Hannibal, was the first Count of Embs, and his descendants often figure in records of the wars of the Austrian Empire, particularly in those connected with the famous Schmalkaldischer Krieg, and are now merged in the family of Count Harrach.

The 'Swiss embroidery' industry here crosses the Rhine, and, in the female gatherings which it occasions, as in the 'Filo' of the south, many local chronicles and legends are, or at least have been, perpetuated.

In the parish church, I have been told by a traveller, that the cardinal's hat of S. Charles Borromeo is preserved, though why it should be so I cannot tell; and I think I have myself had it shown me both at Milan and, if I mistake not, also at the church in Rome whence he had his 'title.'

The ascent to Neu Hohenembs has sufficient difficulty and danger for the unpractised pedestrian to give it special interest, which the roaring of the waterfall tends to excite. A little way beyond it the water was formerly turned to the purpose of an Italian *pescheria* (or fish-preserve for the use of the castle), which is not now very well preserved. Further up still are the ruins of Alt Hohenembs. There are also prettily situated sulphur baths a little way out of the town, much frequented from June to September by the country people. It is curious that the Jews, who have never hitherto settled in large numbers in any part of Tirol, have here a synagogue; and I am told that it serves for nearly a hundred families scattered over the surrounding country, though there are not a dozen even at Innsbruck.

All I have met with of interest between this and Feldkirch, I have mentioned under the head of excursions from Feldkirch.

Stretching along the bank of the Rhine to the south of Feldkirch, is the little principality of Lichtenstein or Liechtenstein, a territory of some three square miles and a half in extent, which yet gives its possessor—lately by marriage made a member of English society—certain seignorial rights. The chief industry of the people is the Swiss embroidery. Vaduz, its chief town, is situated in its centre, and above it, in the midst of a thick wood, is the somewhat imposing and well kept up castle of Lichtenstein. Further south, overhanging the Rhine, is Schloss Gutenberg, and beyond, a remarkable warm sulphur spring, which runs only in summer, at a temperature of 98° to 100° Fahrenheit; it is crowded by Swiss and Tiroleans from June to September, though unknown to the rest of the world.[1] It was discovered in the year 1240 by a chamois-hunter, and was soon after taken in charge by a colony of Benedictine monks, established close by at Pfäffers, who continued to entertain those who visited it until it was taken possession of by the Communal Council of Chur, and the monastery turned into a poor-house. The country round it is exceedingly wild and romantic, and there is a celebrated ravine called the Tamina-Schlund, of so-called immeasurable depth, where at certain hours of a sunny day a wonderful play of light is to be observed. Pfäffers is just outside the boundary of Tirol; the actual boundary line is formed by the Rhætian Alps, which are traversed by a pass called Luziensteig, after St. Lucius, 'first Christian king of Britain,' who, tradition says,

---

[1] It may also be reached by railway as it is but three or four miles from Ragatz, two stations beyond Buchs (p. 13).

preached the gospel to Lichtenstein.[1] The road from Feldkirch to Innsbruck first runs along the Illthal, which between Feldkirch and Bludenz is also called the Wallgau, and merges at Bludenz into the Walserthal on the left or north side. On the right or south side are the Montafonthal, Klosterthal, and Silberthal.

Soon after leaving Feldkirch the mountains narrow upon the road, which crosses the Ill at Felsenau, forming what is called the gorge of the Ill, near Frastanz. Round this terrible pass linger memories of one of the direst struggles for independence the Tiroleans ever waged. In 1499 the Swiss hosts were shown the inlet, through the mountains that so well protect Tirol, by a treacherous peasant whom their gold had bought.[2] A little shepherd lad seeing them advance, in his burning desire to save his country, blew such a call to arms upon his horn that he never desisted till he had blown all the breath out of his little body. The subsequent battle was fierce and determined; and when it slackened from loss of men, the women rushed in and fought with the bravest. So earnestly was the cause of those who fell felt to be the cause of all, that even to the present time the souls of those who were slain that day are remembered in the prayers said as the procession nears the spot

[1] It has been suggested by an eminent comparative mythologist that it is natural *Luc*-ius should be said to have brought 'the Light of the Gospel' to men of *Licht*-enstein.

[2] The traitor was loaded with heavy armour and thrown over the Ill precipice. See Vonbun's parallel with the tradition of the Tarpeian rock, p. 99 n. 2.

when blessing the fields on Rogation-Wednesday. On the heights above Valduna are the striking ruins of a convent of Poor Clares, one of those abandoned at the fiat of Joseph II. It was founded on occasion of a hermit declaring he had often seen a beautiful angel sitting and singing enchantingly on the peak. Below is a tiny lake, which lends an additional charm to the tranquil beauty of the spot. The patron saint of the Walserthal is St. Joder or Theodul (local renderings of Theodoric), and his legend is most fantastic. St. Joder went to Rome to see the Pope; the Pope, in commendation of his zeal, gave him a fine bell for his church. Homewards went St. Joder with his bell, but when he came to the mountains it was more than he could manage, to drag the bell after him. What did he then do? He bethought him that he had, by his prayers and exorcisms, conjured the devil out of the valley where he had preached the faith, so why should not prayer and exorcism conjure him to carry the bell for the service of his faithful flock? If St. Joder's faith did not remove mountains it removed the obstacles they presented, and many a bit of rude carving in mountain chapels throughout the Walserthal shows a youthful saint, in rich episcopal vestments, leading by a chain, like a showman his bear, the arch enemy of souls, crouched and sweating under the weight of the bell whose holy tones are to sound his own ban.[1]

Bludenz retains some picturesque remnants of its

---

[1] Notably at Raggal, Sonntag, Damüls, Luterns, and also in Lichtenstein.—Vonbun, pp. 107-8.

old buildings. It belonged to the Counts of Sonnenberg, and hence it is said that it is often called by that name; but it is perhaps more probable that the height above Bludenz was called Sonnenberg, in contrast with Schattenberg, above Feldkirch, and that its lords derived their name from it. The story of the fidelity of Bludenz to Friedrich *mit der leeren Tasche*, I have narrated in another place.[1]

The valley of Montafon has for its arms the cross keys of St. Peter, in memory of a traditionary but anachronistic journey of Pope John XXIII. to the Council of Constance, in 1414.[2] In memory of the same journey a joy-peal is rung on every Wednesday throughout the year.

A little way south of Bludenz, down the Montafon valley, is a chapel on a little height called S. Anton, covering the spot where tradition says was once a mighty city called Prazalanz, destroyed by an avalanche. Near here is a tiny stream, of which the peasants tell the following story:—They say up the mountain lives a beautiful maiden, set to guard a treasure, and she can only be released when some one will thrice kiss a loathsome toad,[3] which has its place on the cover of the treasury, and the maiden feels assured no one will ever make the venture. She weeps

---

[1] *Infra*, Chapter viii., p. 238.
[2] Vonbun, pp. 92-3.
[3] Some analogous cases quoted in *Sagas from the Far East*, pp. 365, 383-5.

evermore, and they call this streamlet the 'Trächnabüchle'—the Tear-rill.

The valley of Montafon is further celebrated for its production of kirschwasser.

Opposite Dalaas is a striking peak, attaining an elevation of some 5,000 feet, called the Christberg. On the opposite side to Dalaas is a chapel of St. Agatha; in the days of the silver mining of Tirol, in the fifteenth century, silver was found in this neighbourhood. On one occasion a landslip imprisoned a number of miners in their workings. In terror at their threatened death, they vowed that if help reached them in time, they would build a chapel on the spot to commemorate their deliverance. Help did reach them, and they kept their vow. The chapel is built into the living rock where this occurred, and a grey mark on the rock is pointed out as a supernatural token which cannot be effaced, to remind the people of the deliverance that took place there. It is reached from Dalaas by a terribly steep and rugged path, running over the Christberg, near the summit of which may be found, by those whom its hardships do not deter, another chapel, or wayside shrine, consisting of an image of the Blessed Virgin under a canopy, with an alcoved seat beneath it for the votary to rest in, called 'Das Bruederhüsle,' and this is the reason of its name :—The wife of a Count Tanberg gave birth to a dead child; in the fulness of their faith, the parents mourned that to the soul of their little one Christian baptism had been denied, more than the loss of their offspring. In pursuance of a custom then in vogue in parts of Tirol, if not elsewhere, the Count sent the body of the infant to be laid on the altar of St. Joseph, in the parish church, in the hope that at the intercession of the fosterfather of the Saviour it might revive for a sufficient interval to receive the sacrament of admission into the Christian family. The servant, however, instead of carrying his burden to the church at Schruns (in Montafonthal), finding himself weary by the time he had climbed up the Christberg, dug a grave, and buried it instead. The next year there was another infant, also born dead; this time the Count determined to carry it himself to the church, and by the time he had toiled to the same spot he too was weary, and sat down to rest. As he sat he heard a little voice crying from under the ground, '*ätti, nüm mi' ó met!*' [1] The Count turned up the soil, and found the body of his last year's infant. Full of joy he carried both brothers to the altar of St. Joseph, at Schruns; here, continues the legend, his prayer went up before the divine throne; both infants gave signs of life before devout witnesses; baptism could be validly administered, and they, laid to rest in holy ground.[2]

After Dalaas the road assumes a character of real grandeur, both as an engineering work and as a study of nature. The size of the telegraph poles alone (something like fourteen inches in diameter) gives an

---

[1] Father! take me also with you.
[2] Vonbun, pp. 115-7.

idea of the sort of storms the road is built to resist ; so do the veritable fortifications, erected here and there, to protect it from avalanches.

The summit (6,218 ft.) of the Arlberg, whence the province has its name—and which in turn is named from Schloss Arlen, the ruins of which are to be observed from the road—is marked by a gigantic crucifix, overhanging the road. An inscription cut in the rock records that it was opened for traffic (after three hard years of labour) on St. James's day, 1787; but a considerable stretch of the road now used was made along a safer and more sheltered pass in 1822-4, when a remarkable viaduct called the Franzensbrücke was built. Two posts, striped with the local colours, near the crucifix above-named, mark the boundary of Vorarlberg and Oberinnthal. As we pass them we should take leave of Vorarlberg; but it may be convenient to mention in this place some few of the more salient of the many points of interest on the onward road to Innsbruck.

The opening of the Stanzerthal, indeed, on which the road is carried, seems to belong of right to Vorarlberg, for its first post-halt of S. Christof came into existence through the agency of a poor foundling boy of that province, who was so moved by the sufferings of travellers at his date (1386), that he devoted his life to their service, and by begging collected money to found the nucleus of the hospice and brotherhood of S. Christof, which lasted till the time of Joseph II. The pass at its highest part is free from snow only from the beginning of July to September, and in the depth of winter it accumulates to a height of twenty feet. The church contains considerable remains of the date of its founder, *Heinrich das Findelkind*; of this date, or not much later, must be the gigantic statue of S. Christopher, patron of wayfarers.

The Stanzerthal, without being less grand, presents a much more smiling prospect than that traversed during the later part of the journey through Vorarlberg. The waters of the Rosanna and the Trisanna flow by the way; the mountains stretch away in the distance, in every hue of brilliant colouring; the whole landscape is studded with villages clustering round their church steeples, while Indian-corn-fields, fruit-gardens in which the barberry holds no insignificant place, and vast patches of a deep-tinted wild flora, fill up the picture.

At Schloss Wiesburg is the opening into the Patznaunthal, the chief village of which is Ischgl, where the custom I have heard of in other parts of Tirol, and also in Brittany, prevails, of preserving the skulls of the dead in an open vault in the churchyard, with their names painted on them. Nearly opposite it, off the left side of the road lies Grüns or Grins, so called because it affords a bright green patch amid the grey of the rocks. It was a more important place in mediæval times, for the road then ran beside it; the bridge with its pointed arches dates from the year 1639. Margareta Maultasch, with whose place in Tirolese history we must make acquaintance further on, had a house here which still contains some curious mural paintings.

Landeck [1] is an important thriving little town, with the Inn flowing through its midst. It has two fine remains of ancient castles: Schloss Landeck, now used partly as a hospice; and Schloss Schrofenstein, of difficult access, haunted by a knight, who gave too ready ear to the calumnies of a rejected suitor of his wife, and must wander round its precincts wringing his fettered hands and crying 'Woe!' On the slope of the hill crowned by Schloss Landeck stands the parish church. Its first foundation dates from the fifteenth century, when a Landecker named Henry and his wife Eva, having lost their two children in a forest, on vowing a church in honour of the Blessed Virgin, met a bear and a wolf each carrying one of the children tenderly on its back. It has a double-bulbed tower of much later date, and it was restored with considerable care a few years back; but many important parts remain in their original condition, including some early sculpture. In the churchyard are two important monuments, one dating from the fifteenth century, of Oswald Y. Schrofenstein; the other, a little gothic chapel, consecrated on August 22, 1870, in memory of the Lan-

---

[1] The story of its curious success against the Bavarians in 1703, p. 287-8. From Landeck there is a fine road (the description of which belongs to Snitt-Tirol), over the Finstermünz and Stelvio, to the baths of Bormio or Worms.

deck contingent of the Tirolean sharpshooters, who assisted in defending the borders of Wälsch-Tirol in 1866.[1] About two or three miles from Landeck there is a celebrated waterfall, at a spot called Letz.

Imst was formerly celebrated for its breed of canary-birds, which its townsmen used to carry all over Europe. The church contains a votive tablet, put up by some of them on occasion of being saved from shipwreck in the Mediterranean. It has a good old inn, once a knightly palace. From Imst the Pitzthal branches southwards; but concerning it I have not space to enlarge, as the more interesting excursion to Füssen, on the Baravian frontier, must not be passed over. The pleasantest way of making this excursion is to engage a carriage for the whole distance at Imst, but a diligence or 'Eilwagen,' running daily between Innsbruck and Füssen, may be met at Nassereit, some three miles along the Gunglthal. At Nassereit I will pause a moment to mention a circumstance, bearing on the question of the formation of legends, which seemed to take considerable hold on the people, and was narrated to me with a manifest impression of belief in the supernatural. There was a pilgrimage from a place called Biberwier to a shrine of the Virgin, at Dormiz, on August 10, 1869. It was to gain the indulgence of the Vatican Council,

---

[1] The chief encounter occurred at a place called Le Tezze, near Primolano, on the Venetian border, where the Tiroleans repulsed the Italians, in numbers tenfold greater than their own, and no further attempt was made. The anniversary is regularly observed by visiting the graves on August 14; mentioned below at Le Tezze.

and the priest of Biberwier in exhorting his people to treat it entirely as a matter of penance, and not as a party of pleasure, had made use of a figure of speech bidding them not to trust themselves to the bark of worldly pleasure, for, he assured them, it had many holes in it, and would swamp them instead of bearing them on to the joys of heaven. Four of the men, however, persisted in disregarding his warning, and in combining a trip to the Fernsee, one of two romantically situated mountain lakes overlooked by the ancient castle of Sigmundsburg, on a promontory running into it and with its Wirthshaus 'auf dem Fern' forming a favourite though difficult pleasure-excursion. The weather was treacherous; the boat was swamped in the squall which ensued, and all four men were drowned. From Nassereit also is generally made the ascent of the Tschirgants, the peak which has constantly formed a remarkable feature in the landscape all the way from Arlberg.

The road to Füssen passes by Sigmundsburg, Fernsee and Biberwier mentioned in the preceding narrative also the beautiful Blendsee and Mittersee (accessible only to the pedestrian) or rather the by-paths leading to them. Leermoos is the next place passed,—a straggling, inconsiderable hamlet, but affording a pleasing incident in the landscape, when, after passing it, the steep road winds back upon it and reveals it again far far below you. It is, however, quite possible to put up for a night with the accommodation afforded by the Post inn, and by this means one of the most justly celebrated natural beauties may be enjoyed, in the sunset effects produced by the lighting up of the Zugspitzwand.

Next is Lähn, whose situation disposes one to believe the tradition that it has its name from the avalanches (*Lawinen*, locally contracted into Lähne) by which the valley is frequently visited, and chiefly from a terrible one, in the fifteenth century, which destroyed the village, till then called Mitterwald. A carrier who had been wont to pass that way, struck with compassion at the desolation of the place, aided in providing the surviving inhabitants to rebuild their chapel, and tradition fables of him that they were aided by an angel. The road opens out once more as we approach Heiterwang; there is also a post-road hence to Ammergau; here, a small party may put up at the Rossl, for the sake of visiting the Plansee, the second largest lake of Tirol, on the right (east) of the road; on the left is the opening of the Lechthal, a difficult excursion even to the most practised pedestrian. For those who study convenience the Plansee may be better visited from Reutte.

After Heiterwang the rocks close in again on the road as we pass through the Ehrenberger Klause, celebrated again and again through the pages of Tirolese history, from the very earliest times, for heroic defences; its castle is an important and beautiful ruin; and so the road proceeds to Reutte, Füssen, and the much visited Lustschloss of Schwangau; but as these are in Bavaria I must not occupy my Tirolese pages with them, but mention only the Mangtritt, the boundary pass, where a cross stands out boldly against the sky, in memory of S. Magnus, the apostle of these valleys. The devil, furious at the success of the saint with his conversion of the heathen inhabitants, sent a tribe of wild and evil men, says one version of the legend, a formidable dragon acording to another, to exterminate him; he was thus driven to the narrow glen where the fine post-road now runs between the rocks beside the roaring Lech. Nothing daunted, the saint sprang across to the opposite rock whither his adversaries, who had no guardian angels' wings to 'bear them up', durst not pursue him; it is a curious fact for the comparative mythologist that the same pass bears also the name of Jusulte (Saltus Julii) and the tradition that Julius Cæsar performed a similar feat here on horseback. Near it is a poor little inn, called 'the White House,' where local vintages may be tasted.

Reutte has two inns; the *Post and Krone*, and from it more excursions may be made than I have space to chronicle. That to Breitenerang is an easy one; a house here is pointed out as having been built on the spot where stood a poor hut which gave shelter in his last moments to Lothair II. 'the Saxon' overtaken by death on his return journey from the war in Italy, 1137; what remained of the old materials having been conscientiously worked into the building, down to the most insignificant spar; a tablet records the event. The church, a Benedictine foundation of the twelfth century, was rebuilt in the seventeenth, and contains many specimens of what Tirolese artists can do in sculpture, wood-carving, and painting. A quaint chapel in the church-yard has a representation in stucco of the 'Dance of Death.'

The country between this and the Plansee is called the Achenthal, fortunately distinguished by local mis-pronounciation as the Archenthal from the better known (though not deservingly so) Achenthal, which we shall visit later. The Ache or Arche affords several water-falls, the most important of them, the Stuibfall, is nearly a hundred feet in height, and on a bright evening a beautiful 'iris' may be seen enthroned in its foam.

At the easternmost extremity of the Plansee, to be reached either by pleasure boat or mountain path, near the little border custom-house, the Kaiser-brunnen flows into the lake, so called because its cool waters once afforded a refreshing drink to Ludwig of Brandenberg, when out hunting: a crucifix marks the spot. There is also a chapel erected at the end of the 17th century, in consequence of some local vow, containing a picture of the 'Vierzehn Nothhelfer;' and as the so-called 'Fourteen Helpers in Need' are a favourite devotion all over North-Tirol I may as well mention their legend here at our first time of meeting them. The story is that on the feast of the Invention of the Cross, 1445, a shepherd-boy named Hermann, serving the Cistercian monks of Langheim (some thirty miles south of Mayence) was keeping sheep on a farm belonging to them in Frankenthal not far from Würtzburg, when he heard a child's voice crying to him out of the long grass; he turned round and saw a beautiful infant with two tapers burning before it, who disappeared as he approached. On the vigil of S. Peter in the following year Hermann saw the same vision repeated, only this time the beautiful infant was surrounded by a court of fourteen other children, who told him they were the 'Vierzehn Nothelfer,' and that he was to build a chapel to them. The monks refused to believe Hermann's story, but the popular mind connected it with a devotion which was already widespread, and by the year 1448 the mysteriously ordered chapel was raised, and speedily became a place of pilgrimage. This chapel has been constantly maintained and enlarged and has now grown into a considerable church; and the devotion to the 'Fourteen Helpers in Need' spread over the surrounding country with the usual rapid spread of a popular devotion.[1]

[1] Following are the names of the fourteen, but I have never met any one who could explain the selection. 1. S. Acatius, bishop in Asia Minor, saved from death in the persecutions under Decius, 250, by a miracle he performed in the judgment hall where he was tried, and in memory of which he carries a tree, or a branch of one, in pictures of him. 2. S. Ægidius (Giles, in German, Gilgen), Hermit, of Nimes, nourished in his cell by the milk of a hind, which, being hunted, led to the discovery of his sanctity, an episode constantly recurring in the legendary world. Another poetical legend concerning him is that a monk, having come to him to express a doubt as to the virginity of Our Lady, S. Giles, for all answer traced her name in the sand with his staff, and forthwith full-bloom lilies sprang up out of it. 3. S. Barbara. A maiden whom her heathen father shut up in a tower, that nothing might distract her attention from the life of study to which he devoted her; among the learned men who came to enjoy her elevated conversation came a Christian teacher, and converted her; in token of her belief in the doctrine of the Trinity she had three windows made

THE VALLEYS OF TIROL.

The chief remaining points of interest in the further journey to Innsbruck, taking it up where we diverged from it at Nassereit, are mentioned later in my excursions for Innsbruck.

in her tower, and by the token her father discovered her conversion, delivered her to judgment, and she suffered an incredible repetition of martyrdoms. She is generally painted with her three-windowed tower in her hand. 4. S. Blase, Bishop of Sebaste and Martyr, A.D. 288. He had studied medicine, and when concealed in the woods during time of persecution, the wild beasts used to bring the wounded of their number to his feet to be healed. Men hunting for Christians to drag to justice, found him surrounded by lions, tigers, and bears; even in prison he continued to exercise his healing powers, and from restoring to life a boy who had been suffocated by swallowing a fishbone, he is invoked as patron against sore throat. He too suffered numerous martyrdoms. 5. S. Christopher. 6. S. Cyriacus, Martyr, 309, concerning whom many legends are told of his having delivered two princesses from incurable maladies. 7. S. Dionysius, the Areopagite, converted by S. Paul, and consecrated by him Bishop of Athens, afterwards called to Rome by S. Peter, and made Bishop of Paris. 8. S. Erasmus, a bishop in Syria, after enduring many tortures there, he was thrown into prison, and delivered by an angel, who sent him to preach Christianity in Italy, he died at Gaeta 303. At Naples and other places he is honoured as S. Elmo. 9. S. Eustachius, originally called Placidus, a Roman officer, converted while hunting by meeting a stag which carried a refulgent cross between its horns; his subsequent reverses, his loss of wife and children, the wonderful meeting with them again, and the agency of animals throughout, make his one of the most romantic of legends. 10. S. George. 11. S. Catherine of Alexandria. 12. S. Margaret. 13. S. Pantaleone, another student of medicine; when, after many tortures, he was finally beheaded, the legend tells us that, in token of the purity of his life, milk flowed from his veins instead of blood, A.D. 380. 14. S. Vitus, a Sicilian, instructed by a slave, who was his nurse, in the Christian faith in his early years; his father's endeavours to root out his belief were unavailing, and he suffered A.D. 303, at not more than twelve years of age. The only link I can discover in this chain of saints is that they are all but one or two, whose alleged end I do not know, as S. Christopher, credited with having suffered a plurality of terrible martyrdoms. To each is of course ascribed the patronage over some special one of the various phases of human suffering.

Before closing my chapter on Vorarlberg I must put on record, as a warning to those who may choose to thread its pleasant valleys, a laughable incident which cut short my first attempt to penetrate into Tirol by its means. Our line of route I have already named.[1] Our start was in the most genial of August weather; our party not only harmonious, but humorously inclined; all our stages were full of interest and pleasure, and their memory glances at me reproachfully as I pass them over in rigid obedience to the duty of adhering to my programme. But no, I *must* devote a word of gratitude to the friendly Swiss people, and their kindly hospitable manners on all occasions. The pretty bathing establishments on the lakes, where the little girls go in on their way to school, and swim about as elegantly as if the water were their natural element; the wonderful roofs of Aarau; its late-flowering pomegranates; and the clear delicious water, tumbling along its narrow bed down the centre of all the streets, where we stop to taste of the crystal brook, using the hollow of our hands, pilgrim fashion, and the kind people more than once come out of their houses to offer us glasses and chairs!

I *must* bestow, too, another line of record on the charming village of Rorschach, the little colony of Catholics in the midst of a Protestant canton. Its delicious situation on the Boden-see; our row over the lake by moonlight, where we are nearly run down by one of the steamers perpetually crossing it in all

[1] P. 12.

directions, while our old boatman pours out and loses himself in the mazes of his legendary lore ; the strange effect of interlacing moonbeams, interspersed by golden rays from the sanct lamps with Turner-like effect, seen through the open grated door of the church ; the grotesque draped skeletons supporting the roof of one of the chapels, Caryatid fashion and the rustic procession on the early morning of the Assumption.

So far all had gone passing well ; my first misgiving arose when I saw the factotum of the Oberriet station eye our luggage, the provision of four English winterers in Rome, and a look of embarrassed astonishment dilate his stolid German countenance. It was evident that when he engaged himself as ticket-clerk, porter, 'and *everyting*,' he never contemplated such a pile of boxes being ever deposited at his station. We left him wrapt in his earnest gaze, and walked on to see what help we could get in the village. It was a collection of a half-dozen cottages, picturesque in their utter uncivilization, clustered round an inn of some pretensions. The host had apparently heard of the depth of English purses, and was delighted to make his *premières armes* in testing their capacity. Of course there was ' no arguing with the master of' the only horses to whose assistance we had to look for carrying us beyond the mountains, which now somehow struck us as much more plainly marked on the map than we had noticed before. His price had to be ours, and his statement of the distance, about double the reality, had to be accepted also. His stud was soon displayed before us. Three rather tired greys were brought in from the field, and made fast (or rather loose) with ropes to a waggon, on which our formidable *Gepäck* was piled, and took their start with funeral solemnity. An hour later a parcel of boys had succeeded in capturing a wild colt destined to assist his venerable parent in transporting ourselves in a 'shay,' of the Gilpin type, and to which we managed to hang on with some difficulty, the wild-looking driver good-naturedly volunteering to run by the side.

Off we started with the inevitable thunder of German whip-cracking and German imprecations on the cattle, sufficient for the first twenty paces to astonish the colt into propriety. No sooner had we reached the village boundary, however, than he seemed to guess for the first time that he had been entrapped into bondage. With refreshing juvenile buoyancy he instantly determined to show us his indomitable spirit. Resisting all efforts of his companion in harness to proceed, he suddenly made such desperate assault and battery with his hind legs, that one or two of the ropes were quickly snapped, the Jehu sent sprawling in the ditch on one side, and the travelling bags on the other; so that, but for the staid demeanour of the old mare, we should probably in two minutes more have been 'nowhere.' Hans was on his feet again in an instant, like the balanced mannikins of a bull-fight, and to knot the ropes and make a fresh start required only a minute more; but another and another exhibition of the colt's pranks decided us to trust to our own powers of locomotion.

A bare-footed, short-petticoated wench, who astonished us by proving that her rough hands could earn her livelihoood at delicate 'Swiss' embroidery, and still more by details of the small remuneration that contented her, volunteered to pilot us through the woods where we had quite lost our way; and finding our luggage van waiting on the banks of the Rhine for the return of the ferry, we crossed with it and walked by its side for the rest of the distance.

Our road lay right across the Ardetzen, a basin of pasture enclosed by a magnificent circuit of mountains,— behind us the distant eminences of Appenzell, before us the great Rhætian Alps, and at their base a number of smiling villages each with its green spire scarcely detaching from the verdant slopes behind. The undertaking, pleasant and bright at first, grew weary and anxious as the sun descended, and the mountains of Appenzell began to throw their long shadow over the lowland we were traversing, and yet the end was not reached. At last the strains of an organ burst upon our ears, lights from latticed windows diapered our path, and a train of worshippers poured past us to join in the melodies of the Church, sufficiently large to argue that our stopping-place was attained. We cast about to find the *Gasthof zur Post* to which we were bound, but all in vain, there was no rest for us.

Here indeed, Feldkirch *fuit,* but here it was no more. In the year 909, the Counts of Montfort built themselves a castle on the neighbouring height of Schattenburg, (so called because the higher eminences around shade it from the sun till late in the morning,) and lured away the people from this pristine Feldkirch to settle themselves round the foot of their fortress. Some of the original inhabitants still clung to the old place, and its old Church of St. Peter, that very church whose earlier foundations, some say, were laid by monks from Britain, S. Columban and St. Gall, who, when the people were oppressed by their Frankish masters, came and lived among them, and by their preaching and their prayers rekindled the light of religion, working out at the same time their political relief; the former subsequently made his way, shedding blessings as he went, on to Italy, where he died at the age of ninety, in 615; the latter founded, and ended his days at the age of ninety-five, in the famous monastery which has given his name to the neighbouring Swiss Canton.

The descendants of this remnant have kept up the original settlement to this day with the name of Altenstadt, while the first built street of the present thriving town of Feldkirch still retains its appellation of the Neustadt.

It seemed a long stretch ere we again came upon an inhabited spot, but this time there was no mistake. All around were the signs of a prosperous centre, the causeways correctly laid out, new buildings rising on every side, and—I am fain to add—the church dark and closed; in place of the train of worshippers of unsophisticated Altenstadt, one solitary figure in mourning weeds was kneeling in the moonlight at a desk such as we often see placed under a cross against the outer wall of churches in Germany.

Before five next morning I was awakened by the pealing organ and hearty voices of the Feldkirch peasants at Mass in the church just opposite my window. I dressed hastily, and descended to take my place among them. It was a village festival and Mass succeeded Mass at each of the gaily decorated altars, and before them assembled groups in quaint costumes from far and near.[1] As each half hour struck, a bell sounded, and a relic was brought round to the high altar rails, all the women in the church going up first, and then all the men, to venerate it.

Our first care of the day was to engage our carriage for Innsbruck. We were at the Post hotel, and had the best chance there; for besides its own conveyances, there were those of the post-office, which generally in Germany afford great convenience. Not one was there,

[1] Among these not the least remarkable were some specimens of the unbrimmed beaver hat, somewhat resembling the Grenadier's bearskin, only shorter, which is worn by the women in various parts of Tirol and Styria.

however, that would undertake our luggage over the mountain roads. The post-master and his men all declared that at every winding of the passes there would be too great risk of overturning the vehicle. It was in vain we argued that the same amount had often accompanied us over higher mountains in Italy; it was clear they were not prepared for it. There was a service for heavy goods by which it could be sent; there was no other way, and they did not advise that. They could not ensure any due care being taken of it, or that it should reach within three or four weeks. Four or five hours spent in weighing, measuring, arranging, and arguing, advanced our cause not a whit; there was no plan to be adopted but to return by Oberriet to Rorschach, cross lake Constance to Lindau, and make our way round by Augsburg, Munich, and Rosenheim!

It was with great reluctance we relinquished the cherished project. Our now hated luggage deposited in a waggon, as the day before, we mounted our rather more presentable, and certainly better horsed vehicle, in no cheerful mood, for, besides the disappointment, there was the mortification which always attaches to a failed project and retraced steps.

'The *Herrschaften* are not in such bright spirits as the sun to-day!' exclaimed our driver, when, finally tired of cracking his whip and shouting to his horses, he found we still sat silent and crest-fallen. He wore the jauntiest costume to be found in Europe, after that of his Hungarian confrère, a short postilion jacket, bound and trimmed with yellow lace, a horn slung across his breast by a bright yellow cord, and a hat shining like looking-glass cocked on one side of his head, while his face expressed everything that is pleasant and jovial.

'How can one be anything but out of spirits when one is crossed by such a stupid set as the people of your town? Why, there is no part of Europe in which they will even believe it possible!'

'Well, you see they *don't* understand much, about here,' he replied, with an air of superiority, for he was a travelled postilion, as he took care to let us know. 'In Italy they manage better; they tie the luggage on behind, or underneath, where it is safe enough. Here they have only one idea—to stick it on the top, and in that way a carriage may be easily upset at a sharp turn. You cannot drive any new idea into these fellows; it is like an echo between their own mountains, whatever is once there, goes on and on and on.' I showed him the map, and traced before him the difference in the length of the route we should have taken and that we had now to pursue. I don't think he had ever understood a map before, for he seemed vastly pleased at the compliment paid to his intelligence. 'Ah!' he exclaimed, 'if we could always go as the crow flies, how quickly we should get to our journey's end; or if we had the Stase-Sattel, as they used to have—wasn't *that* fine!'

'The Stase-Sattel,' I replied, 'what is that?'

'What! don't you know about the Stase-Sattel—at that place, Bludenz, there,' and he pointed to it on the map,' 'where you were telling me you wanted to have gone, there used to live an old woman named Stase, and folk said she was a witch. She had a wonderful saddle, on to which she used to set herself when she wanted anything, and it used to fly with her ever so high, and quicker than a bird. One day the reapers were in a field cooking their mess, and they had forgotten to bring any salt—and *hupf!* quick! before the pot had begun to boil she had flown off on her saddle to the salt-mines at Hall, beyond Innsbruck, and back with salt enough to pickle an ox. Another time there was a farmer who had been kind to her, whose crops were failing for the drought. She no sooner heard of his distress than up she flew in her saddle and swept all the clouds together with her broom till there was enough to make a good rainfall. Another time, a boy who had been sent with a message by his master to the next village had wasted all the day in playing and drinking with her; towards dusk he bethought himself that the gates would be shut and the dogs let loose, so that it was a chance if he reached the house alive. But she told him not to mind, and taking him up on her saddle, she carried him up through the air and set him down at home before the sun was an inch lower.'

'And what became of her?' I inquired.

'Became of her! why, she went the way of all such folk. They go on for a time, but God's hand overtakes them at the last. One day she was on one of her wild errands, and it was a *Fest-tag* to boot. Her course took her exactly over a church spire, and just as she passed, the *Wandlung* bell[1] tolled. The sacred sound tormented her so that she lost her seat and fell headlong to the ground. When they came out of church they found her lying a shapeless mass upon the stone step of the churchyard cross. Her enchanted saddle was long kept in the Castle of Landeck—maybe it is there yet; and even now when we want to tell one to go quickly on an errand, we say, " Fly on the saddle of Dame Stase." '

'You have had many such folk about here,' I observed seriously, with the view of drawing him out.

'Well, yes, they tell many such tales,' he answered; 'and if they're not true, they at least serve to keep alive the faith that God is over us all, and that the evil one has no more power than just what He allows. There's another story they tell, just showing that,' he continued. ' Many years ago there was a peasant (and he lived near Bludenz too) who had a great desire to have a fine large farm-house. He worked hard, and put his savings by prudently; but it wouldn't do, he never could get enough. One day, in an evil hour, he let his great desire get the better of him, and he called

---

[1] The bell called in other countries the *Elevation bell*, is in Germany called the *Wandlung*, or *change-of-the-elements* bell. The idiom was worth preserving here, as it depicts more perfectly the solemnity of the moment indicated.

the devil in *dreiteufelsnamen*[1] to his assistance. It was not, you see, a deliberate wickedness—it was all in a moment, like. But the devil came, and didn't give him time to reflect. "I know what you want," he said; "you shall have your house and your barns and your hen-house, and all complete, this very night, without costing you a penny; but when you have enjoyed it long enough, your old worn-out carcass shall belong to me." The good peasant hesitated; and the devil, finding it necessary to add another bait, ran on: "And what is more, I'll go so far as to say that if every stone is not complete by the first cock-crow, I'll strike out even this condition, and you shall have it *out and out*." The peasant was dazzled with the prospect, and could not bring himself all at once to refuse the accomplishment of his darling hope. The devil shook him by the hand as a way of clenching the bargain, and disappeared.

'The peasant went home more alarmed than rejoiced, and full of fear above all that his wife should inquire the meaning of all the hammering and blustering and running hither and thither which was to be heard going on in the homestead, for she was a pious God-fearing woman.

'He remained dumb to all her inquiries, hour after hour through the night; but at last, towards morning, his courage failed him, and he told her all. She, like a good wife, gave back no word of reproach, but cast

---

[1] The threefold invocation, supposed to be supremely efficacious.

about to find a remedy. First she considered that he had done the thing thoughtlessly and rashly, and then she ascertained that at last he had given no actual consent. Finally, deciding matters were not as bad as might be, she got up, and bid him leave the issue to her.

'First she knelt down and commended herself and her undertaking to God and His holy saints; then in the small hours, when the devil's work was nearly finished, she took her lamp and spread out the wick so that it should give its greatest glare, and poured fresh oil upon it, and went out with a basket of grain to feed the hens. The cock, seeing the bright light and the good wife with her basket of food, never doubted but that it was morning, and springing up, he flapped his wings, and crowed with all his might. At that very moment the devil himself was coming by with the last roof-stone.[1] At the sound of the premature cock-crow he was so much astonished that he didn't know which way to turn, and sank into the ground bearing the stone still in his hand.

'The house belonged to the peasant by every right, but no stone could ever be made to stay on the vacant space. This inconvenience was the penance he had to endure for the desperate game he had played, and he

---

[1] In Tirol the roofs are frequently made of narrow overlapping planks, weighed down by large stones. Hence the origin of the German proverb, 'If a stone fall from the roof, ten to one but it lights on a poor widow;'—equivalent to our 'Trouble never comes alone.'

took it cheerfully, and when the rain came in he used to kiss his good wife in gratitude for the more terrible chastisement from which she had saved him.'

The jaunty postilion whipped the horses on as he thus brought his story to a close, or rather cracked his whip in the air till the mountains resounded with it, for he had slackened speed while telling his tale, and the day was wearing on.

'We must take care and not be late for the train,' he observed. 'The *Herrschaften* have had enough of the inn of Oberriet, and don't want to have to spend a night *there*, and we have no *Vorarlberger-geist* to speed us now-a-days.'

'Who was he?' I inquired eagerly.

'I suppose you know that all this country round about here is called the Vorarlberg, and in olden time there was a spirit that used to wander about helping travellers all along its roads. When they were benighted, it used to go before them with a light; when they were in difficulties, it used to procure them aid; if one lost his way, it used to direct him aright; till one day a poor priest came by who had been to administer a distant parishioner. His way had lain now over bog, now over torrent-beds. In the roughness of the way the priest's horse had cast a shoe. A long stretch of road lay yet before him, but no forge was near. Suddenly the *Vorarlberger-geist* came out of a cleft in the rock, silently set to work and shod the horse, and passed on its way as usual with a sigh.

'"*Vergeltsgott!*"[1] cried the priest after it.

"'God be praised!' exclaimed the spirit. "Now am I at last set free. These hundred years have I served mankind thus, and till now no man has performed this act of gratitude, the condition of my release." And since this time it has never been seen again.'

We had now once more reached the banks of the Rhine. The driver of the luggage van held the ferry in expectation of us, and with its team it was already stowed on board. Our horses were next embarked, and then ourselves, as we sat, perched on the carriage. A couple of rough donkeys, a patriarchal goat, and half-a-dozen wild-looking half-clothed peasants, made up a freight which seemed to tax the powers of the crazy barge to the utmost; and as the three brawny ferrymen pulled it dexterously along the guide rope, the waters of the here broad and rapid river rose some inches through the chinks. All went well, however, and in another half-hour we were again astonishing the factotum of the Oberriet station with a vision of the 'Gepack' which had puzzled him so immensely the day before.

[1] 'May God reward it.'

CHAPTER II.

NORTH-TIROL—UNTERINNTHAL (RIGHT-INN BANK).

KUFSTEIN TO ROTTENBURG.

> . . . . '*Peasant of the Alps,*
> *Thy humble virtues, hospitable home,*
> *And spirit patient, pious, proud, and free;*
> *Thy self-respect, grafted on innocent thoughts;*
> *Thy days of health, thy nights of sleep, thy toil*
> *By danger dignified, yet guiltless; hopes*
> *Of cheerful old age, and then a quiet grave*
> *With cross and garland over its green turf,*
> *And thy grandchildren's love for epitaph,*
> *This do I see!* . . . . .
>
> BYRON (*Manfred*).

WHEN, after our forced *détour*, we next penetrated into Tirol, it was by the way of Kufstein. Ruffled as we had been in the meantime by Bavarian '*Rohheit*,' we were glad to find ourselves again in the hands of the gentle Tirolese.

Kufstein, however, is not gentle in appearance. Its vast fortress seems to shed a stifling gloom over the whole place; it looks so hard and selfish and tyrannical, that you long to get away from its influence. Noble hearts from honest Hungary have pined away within its cold strong grasp; and many a time, as my sketch-book has been turned over by Magyar friends, the page which depicted its outline—for it wears a grand

and gallant form, such as the pencil cannot resist—has raised a deep sigh over the '*trauriges Andenken*' it served to call up.[1]

When Margaretha Maultasch ceded the country she found herself unable to govern, to Austria at the earnest request of her people, in 1363, it was stipulated that Kufstein, Kitzbühl, and Rattenberg, which had been added to it by her marriage with Louis of Bradenburg, should revert to Bavaria. These three dependencies were recovered by the Emperor Maximilian in 1504, the two latter accepting his allegiance gladly, the former holding out stoutly against him. The story of the reduction of this stronghold is almost a stain on his otherwise prudent and prosperous reign.

Pienzenau, its commander, who was in the Bavarian interest, had particularly excited his ire by setting his men to sweep away with brooms the traces of the small damage which had been effected by his cannon, placed at too great a distance to do more than graze the massive walls. Philip von Recenau, Regent of Innsbruck, meantime cast two enormous field-pieces, which received the names of *Weckauf* and *Purlepaus*. These entirely turned the tide of affairs. Chronicles of the time do not mention their calibre, but declare that their missiles not only pierced the 'fourteen feet-thick wall' through and through, but entered a foot and a half into the living rock. Pienzenau's heart misgave him when he saw the

---

[1] The frontispiece to this volume (very much improved by the artist who has drawn it on the wood).

work of these destructive engines, and hastened to send in his submission to the Emperor; but it was too late. 'So he is in a hurry to throw away his brooms at last, is he?' cried Maximilian. 'But he should have done it before. He has allowed the wall of this noble castle to be so disgracefully shattered, that he can make no amends but by giving up his own carcass to the same fate.'

No entreaty could move the Emperor from carrying out this chastisement, and some five-and-twenty of the principal men who had held out against him were condemned to be beheaded on the spot. When eleven had fallen before the headsman's sword, Erich, Duke of Brunswick, sickening at the scene of blood, pleaded so earnestly with the Emperor, that he obtained the pardon of the rest. The eleven were buried by the pious country-people in a common grave; and who will may yet tread the ground where their remains rest in a little chapel built over their grave at Ainliff (dialectic for eleven), on the other side of the river Inn.

Its situation near the frontier has made it the scene of other sieges, of which none is more endeared to Tiroleans than that of 1809, when the patriot Speckbacher distinguished himself by many a dauntless deed.

If Kufstein has long had a truce to these stirring memories, many a fantastic story has floated out of it concerning the prisoners harboured there, even of late years. The Hungarian patriot brigand, Rocsla Sandor (Andrew Roshla), who won by his unscrupulous daring quite a legendary place in popular story, was long confined here. He was finally tried and condemned (but I think not executed) at Szeghedin, in July 1870; 454 other persons were included in the same trial, of whom 234 under homicidal charges; 100 homicides were laid to his charge alone, but there is no doubt that his services to the popular cause, at the same time that they condoned some of his excesses, in the popular judgment may have disposed the authorities to exaggerate the charges against him. The whole story is fantastic, and even in Kufstein, where he was almost an alien, there was admiration and sympathy underlying the shudder with which the people spoke of him. A much more interesting and no less romantic narrative, was told me of a Hungarian political prisoner, who formed the solitary instance of an escape from the stony walls of the fortress. His lady-love—and she was a lady by birth—with the heroic instincts of a Hungarian maiden, having with infinite difficulty made out where he was confined, followed him hither in peasant disguise, and with invincible perseverance succeeded, first in engaging herself as servant to the governor and then in conveying every day to her lover, in his soup, a hank of hemp. With this he twisted a rope and got safely away; and this occurred not more than six or seven years ago.

St. Louis's day fell while we were at Kufstein—the name-day of the King of Bavaria; and being the border town, the polite Tiroleans make a complimentary fête of it. There was a grand musical Mass, which the officers from the Bavarian frontier attended, and a modest banquet was offered them after it. The peasants put on their holiday attire—passable enough as far as the men are concerned, but consisting mainly on the women's behalf in an ugly black cloth square-waisted dress, and a black felt broad-brimmed hat, with large gold tassels lying on the brim. After Mass the Bavarian national hymn was sung to the familiar strains of our own.

All seemed gay and glad without. I returned to the primitive rambling inn; everyone was gone to take his or her part in the Kufstein idea of a holiday. There were three entrances, and three staircases; I took a wrong one, and in trying to retrace my steps passed a room through the half-open door of which I heard a sound of moaning, which arrested me. I could not find it in my heart to pass on. I pushed the door gently aside, and discovered a grey-haired old man lying comfortlessly on the bed in a state of torpor. I laid him back in a posture in which he could breathe more freely, opened his collar and gave him air, and with the aid of one or two simple means soon brought him back to consciousness. The room was barely furnished; his luggage was a small bundle tied in a handkerchief, his clothes betokened that he belonged to the respectable of the lower class. I was too desirous to converse with a genuine Tirolean peasant to refuse his invitation to sit down by his side. I had soon learnt his tale, which he seemed not a little pleased to find had an interest for a foreigner.

His lot had been marked by severe trials. In early youth he had been called to lose his parents; in later life, the dear wife who had for a season clothed his home again with brightness and hope. In old age he had had a heavier trial still. His only child, the son whom he had reared in the hope that he would have been the staff of his declining years, whom he had brought up in innocence in childhood, and shielded from knowledge of evil in early youth, had gone from him, and he knew not where to find him. The boy had always had a fancy for a roving adventurous life, but it had been his hope to have kept him always near him, free from the contamination of great cities.

I asked if it was not the custom in these parts for young men to go abroad and seek employment where it was more highly paid, and come back and settle on their earnings. But he shook his head proudly. It was so in Switzerland, it was so in some few valleys of Tirol, and the poor Engadeiners supplied all the cities of Europe with confectioners; but his son had no need to tramp the world in search of fortune. But what had made him most anxious was, that the night before his son left some wild young men had passed through the village. They were bold and uproarious, and his fear was that his boy might have been tempted to join them. He did not know exactly what their game was, but he had an idea they were gathering recruits to join the lawless Garibaldian bands in their attempts upon the Roman frontier. With their designs he was confident his son had no sympathy. If he had stopped to consider them, he would have shrunk from them with horror; and it was his dread that his spirited love of danger and excitement had carried him into a vortex from which he might by-and-by be longing to extricate himself in vain. It was to pray that the lad might be guided aright that he made this pilgrimage up the Thierberg—no easy journey for one of his years. He had come across hill and valley from a village of which I forget the name, but situated near Sterzing.

'But Sterzing itself is a place of pilgrimage,' I said, glad to turn to account my scanty knowledge of the sacred places of the country. 'Why did you come all this way?'

'Indeed is Sterzing,' he replied, 'a place of benedictions. It is the spot where Sterzing, our first hermit, lived, and left his name to our town. But *this* is the spot for those who need penance. There, in that place,' and as I followed the direction of his hand I saw through the low lattice window the lofty elevation of the Thierberg like a phantom tower, enveloped in mist, standing out against the clear sky beyond, and wondered how his palsied limbs had carried him up the steep. 'In that place, in olden time, lived a true penitent. Once it was a lordly castle, and he to whom it belonged was a rich and honoured knight; but on one occasion he forgot his knightly honour, and with false vows led astray an unthinking maiden of the village. Soon, however, the conviction of his sin came back to him clear as the sun's light, and without an hour's hesitation he put it from him. To the girl he made the best amends he could by first leading her to repentance, then procuring her admission to a neighbouring convent. But for him, from that day the lordly castle became as a hermit's cell, the sound of mirth and revelry and of friendly voices was hushed for ever. The memory of his own name even he would have wiped out, and would have men call him only, as they do to the present day, '*der Büsser*'—the Penitent. And so many has his example brought to this shrine in a spirit of compunction, that the Church has endowed it with the indulgence of the Portiuncula.'

What a picture of Tirolese faith it was! Instead of setting in motion the detective police, or the telegraph-wire, or the second column of the 'Times,' this old man had come many miles in the opposite direction from that his child was supposed to have taken, to bring his burden and lay it before a shrine he believed to have been made dear to heaven by tears of penance in another age, and there commend his petition to God that He might bring it to pass, accepting the suffering as a merited chastisement in a spirit of sincere penitence!

He was feeling better, and I rose to go. He pressed my hand in acknowledgment of my sympathy, and I assured him of it. It was not a case for more substantial charity; I had gathered from his recital that he had no lack of wordly means. I only strove at parting to kindle a ray of hope. I said after all it might not be so bad as he imagined; his boy had been well brought up, and might perhaps be trusted to keep out of the way of evil. It was thoughtless of him not to seek his father's blessing and consent to his choice of an adventurous career, but it might be he had feared his opposition, and that he had no unworthy reason for concealing his plans. There was at least as much reason to hope as to despond, and he must look forward to his coming back, true to the instincts of his mountain home, wiser than he had set out.

His pale blue eye glistened, and he gasped like one who had seen a vision. 'Ay! just so! Just so it appeared to me when I was on the Thierberg this morning! And now, in case my weak old heart did not see it clearly enough, God, in His mercy, has sent you to expound the thing more plainly to me. Now I know that I am heard.'

Poor old man! I shuddered lest the hope so strongly entertained should prove delusive in the end. I may never know the result; but I felt that at all events as he was one who took all things at God's hands, nothing could, in one sense, come amiss; and for the present, at least, I saw that he went down to his house comforted.

I strolled along the street, and, possessed with the type of the Tirolean peasant, as I received it from this old man, I conceived a feeling of deeper curiosity for all whom I met by the way. I thought of them as of men for whom an unseen world is a reality; who estimate prayer and sacraments and the intercession of saints above steam-power and electricity. At home one meets with one such now and then, but to be transported into a whole country of them was like waking up from a long sleep to find oneself in the age of St. Francis and St. Dominic.

Whatever faults the Tirolese may have to answer for, they will not arise from religion being put out of sight. No village but has its hillside path marked with 'the Way of the Cross;' no bridge but carries the statue of S. John Nepomucene, the martyr of the Confessional; no fountain but bears the image of the local saint, a model of virtue to the place; no lone path unmarked by its way-side chapel, or its crucifix shielded from the weather by a rustic roof; no house but has its outer walls covered with memories of holy things; no room without its sacred prints and its holywater stoup. The churches are full of little rude pictures, recording scenes in which all the pleasanter events of life are gratefully ascribed to answers to prayer, while many who cannot afford this more elaborate tribute hang up a tablet with the words *Hat geholfen* ('He has helped me'), or more simply still, '*aus Dankbarkeit.*' Longfellow has written something very true and pretty, which I do not remember well enough to quote; but most will call to mind the verses about leaving landmarks, which a weary brother seeing, may take heart again; and it is incalculable how these good people may stir up one another to hope and endurance by such testimonies of their trust in a Providence. Sometimes, again, the little tablets record that such an one has undertaken a journey. '*N. N. reiset nach N.*, pray for him;' and we, who have come so far so easily, smile at the short distance which is thought worthy of this importance. The *Gott segne meine Reise*—'May God bless my journey'—seems to come as naturally to them, however, as 'grace before meat' with us. But most of all, their care is displayed in regard to the dear departed. The spot where an accident deprived one of his life is sacred to all. 'The honourable peasant N. N. was run over here by a heavy waggon:'—'Here was N. N. carried away by the waters of the stream:' with the unfailing adjunct, 'may he rest in peace, let us pray for him;' or sometimes, as if there were no need to address the recommendation to his own neighbours, 'Stranger! pray for him.'

The straggling village on the opposite bank of the Inn is called Zell, though appearing part of Kufstein. It affords the best points for viewing the gloomy old fortress, and itself possesses one or two chapels of some interest. At Kiefersfelden, at a short distance on the Bavarian border, is the so-called Ottokapelle, a Gothic chapel marking the spot where Prince Otho quitted his native soil when called to take possession of the throne of Greece.

Kundl, about an hour from Kufstein, the third station, by rail,[1] though wretchedly provided with accommodation, is the place to stop at to visit the curious and isolated church of *S. Leonhard auf der Wiese* (in the meadow), and it is well worthy of a visit. In the year 1004 a life-sized stone image of St. Leonard was brought by the stream to this spot; 'floating,' the wonder-loving people said, but it may well be believed that some rapid swollen torrent had carried the image away in its wild course from some chapel on a higher level. The people not knowing whence it came, reckoned its advent a miracle, and set it up in the highway, that all who passed might know of it. It was not long before a no less illustrious wayfarer than the Emperor Henry II. came that way, and seeing the uncovered image set up on high, stopped to inquire its history. When he had heard it, he vowed that if his arms were prosperous in Italy he would on his return build the saint an honourable church. Success indeed attended him in the campaign, and he was crowned Emperor at Pavia, but St. Leonard and his vow were alike forgotten. The year 1012 brought him again into Italy through Tirol, and passing the spot where he had registered his vow before, his horse, foaming and stamping, refused to pass the image or carry him further. The circumstance reminded him of his promise, and he at once set to work to carry it out worthily. The church was com-

[1] Of the Brixenthal and the Gebiet der grossen Ache we shall have to speak in a later chapter, in our excursion from Wörgl to Vienna.'

pleted within a few years, but an unhappy accident signalized its completion. A young man who had undertaken to place the ornament on the summit was seized with vertigo in the moment of completing his exploit, and losing his balance was dashed lifeless on to the ground below.[1] His remains were gathered up tenderly by the neighbours, and his skull laid as an offering at the foot of the crucifix on the high altar, where it yet remains. An inscription to the following effect is preserved in the church: 'A.D. 1019 Præsens ecclesia Sti. Leonhardi a sancto Henrico Imperatore exstructa, et anno 1020 a summo Pontifice Benedicto VIII. consecrata est,' though there would not seem to be any other record of the Pope having made the journey. S. Kunigunda, consort of Henry II., bore a great affection to the spot, and often visited it.

The image of St. Leonard now in the church bears the date of 1481, and there is no record of the time when it was substituted for the original.[2] The interior has suffered a great deal during the whitewash period; but some of the original carvings are remarkable, particularly the grotesque creatures displayed on the main columns. On one a doubled-bodied lion is trampling on two dragons; on another a youth stands holding the prophetic roll of the book of revelation, and a hideous

---

[1] The comparative mythologist can perhaps tell us why this story crops up everywhere. I have had occasion to report it from Spain in *Patrañas*. Curious instances in Stöber *Sagen des Elsasses*.

[2] S. Leonard is reckoned the patron of herds. See *Pilger durch Tirol*, p. 247.

symbolical figure, with something of the form of a bear, cowers before him, showing a certain resemblance to the sculptures in the chapel-porch of Castle Tirol. Round the high altar are ten pilasters, each setting forth the figure of a saint, and all various. A great deal of the old work was destroyed, however, when it was rebuilt, about the year 1500.

Between St. Leonhard and Ratfeld runs the Auflängerbründl—so called from the Angerberg, celebrated as itself a very charming excursion from Kundl—a watercourse directed by the side of the road through the charity of the townspeople of Rattenberg and Ratfeld, in the year 1424, with the view that no wayfarer might faint by the way for want of a drink of pure and refreshing water.

Rattenberg is a little town of some importance on account of the copper works in the neighbourhood, but not much frequented by visitors, though it has three passable inns. It is curious that the castle of Rottenburg near Rothholz, though so like in name, has a different derivation, the latter arising from the red earth of the neighbourhood, and the former from an old word *Rat*, meaning 'richness,' and in old documents it is found spelt *Rat in berc* (riches in the mountain). This was the favoured locality of the holy Nothburga's earthly career.

St. Nothburga is eminently characteristic of her country. She was the poorest of village maidens, and yet attained the highest and most lasting veneration of her people by the simple force of virtue. She was born in 1280. The child of pious parents, she drank in their good instructions with an instinctive aptitude. Their lessons of pure and Christian manners seemed as it were to crystallize and model themselves in her conduct; she grew up a living picture of holy counsels. She was scarcely seventeen when the lord of Castle Rottenburg, hearing of her perfect life, desired to have her in his household. Her parents, knowing she could have no better protectors, when they were no more, than their honoured knight Henry of Rottenburg and his good wife Gutta, gladly accepted the proposal. In her new sphere Nothburga showed how well grounded was her virtue. It readily adapted itself to her altered position, and she became as faithful and devoted to her employers as she had been loving and obedient to her parents. In time she was advanced to the highest position of trust in the castle, and the greatest delight of her heart was fulfilled when she was nominated to superintend the distribution of alms to the poor. Her prudence enabled her to distinguish between real and feigned need, and while she delighted in ministering to the one, she was firm in resisting the appeals of the other. Her general uprightness won for her the respect of all with whom she had to do, and she was the general favourite of all classes.

Such bright days could not last; the enemy of God's

---

[1] Anna Maria Taigi, lately beatified in Rome, was also a maid-servant.

saints looked on with envy, and desired to 'sift' her 'as wheat.' The knight's son, Henry VI., in process of time brought home his bride, Ottilia by name; and according to local custom, the older Knight Henry ceded his authority to the young castellan, living himself in comparative retirement. Ottilia was young and thoughtless, and haughty to boot, and it was not without a feeling of bitter resentment that she saw both her husband and his parents looked to Nothburga to supply her deficiencies in the management of the household. She resolved to get rid of the faithful servant, and her fury against her was only increased in proportion as she realized that the perfect uprightness of her conduct rendered it impossible to discover any pretext for dismissing her.

For Nothburga it was a life of daily silent martyrdom. There were a thousand mortifications in her mistress's power to inflict, and she lost no opportunity of annoying her, but never once succeeded in ruffling the gentleness of her spirit. 'My life has been too easy hitherto,' she would say in the stillness of her own heart; 'now I am honoured at last by admission to the way of the Cross.' There was no brightness, no praise, no subsequent hope of distinction, to be derived from her patience; they were stabs in the dark, seen by no human eye, which made her bleed day by day. Yet she would not complain, much less seek to change her service. She said it would have been ungrateful to her first benefactors and employers to leave them, so long as she could spend herself for them, and ungrateful to God to shirk the trial He had lovingly sent her.

A crucial test of her fidelity, however, was at hand. The day came when Knight Henry and Gutta his wife were called to their long rest, and with them the chief protection of Nothburga departed. She was now almost at Ottilia's mercy. One of the first consequences of this change was that she was deprived of her favourite office of relieving the poor; and not only their customary alms were stopped, but their dole of food also; and as a final provocation, she was required to feed the pigs with the broken meat which she had been accustomed to husband for the necessitous.

The good girl's heart bled to see the needy whom she had been wont to relieve turned hungry away. The only means that occurred to her of remedying the evil in some measure, was to deny herself her own food and distribute it among them. Restricting her own diet to bread and water, she saved a little basketful, which she would take down every evening when work was done to the foot of the Leuchtenburg, where the poorest of the castle dependents lived; and the blessing which multiplied the loaves in the wilderness made her scanty savings suffice to feed all who had come to beg of her.

That Nothburga contrived to feed the poor of a whole district, in spite of her orders to the contrary, of course became in time a ground of complaint for Ottilia. She had now a plausible reason for stirring up the

Knight Henry against her. He had always defended her, out of regard for his parents' memory; but coming one evening past the Leuchtenburg, at Ottilia's instigation, he met Nothburga with her little burden, and asked her what she carried.

Here the adversary of the saints had prepared for her a great trial, says the legend. She, in her innocence, told fairly and honestly the import of her errand; but to the Knight's eyes, who had meantime untied her apron, the contents appeared, the legend says, to be wood shavings; and further, putting the wine-flask to his lips, it seemed to him to contain soap-suds. To her charitable intention he had made no objection, but at this, which appeared to him a studied affront, he was furious. He would listen to no explanation, but, returning at once to the castle, he gave Ottilia free and full leave to deal with the offending handmaiden as she pleased. Ottilia readily put the permission into effect by directing the castle guard to forbid her, on her return, ever again to pass the threshold of the castle.

This blow told with terrible effect on the poor girl. During her service at the castle both her parents had died; she had now no home to resort to. Putting her trust in God, however, she retraced her steps alone through the darkness, and found shelter in a cottage of one of her clients. Her path was watched by the angels, who marked the track with fair seeds; and even to this day the hill-side which her feet so often pressed on her holy errand is said to be marked with a peculiar growth of flowers.

The next day she applied to a peasant of Eben to engage her as a field labourer. The peasant was exceedingly doubtful of her capacity for the work after the comparatively delicate nature of her previous mode of life. Her hardy perseverance and determination, aided by the grace of God, on which she implicitly relied, overcame all obstacles, and old Valentine soon found that her presence brought a blessing on all his substance. She had been with him about a year, when one day, being Saturday, he was very anxious to gather in the remainder of his harvest before an apprehended storm, and desired Nothburga, with the other reapers, to continue their labours after the hour of eve, when the holy rest was reckoned to have commenced. Nothburga, usually so obedient to his wishes, had the courage to refuse to infringe the commandment of religion; and to manifest that the will of God was on her side, showed him her sickle resting from labour, suspended in the air. Valentine, convinced by the prodigy, yielded to her representations, and her piety was more and more honoured by all the neighbours.

Soon after this, Ottilia, in the midst of her health and strength, was stricken with a dangerous illness. In presence of the fear of death she remembered her harsh treatment of Nothburga, and sent for her to make amends for the past. As the good girl reached her bed-side she was just under the influence of a frightful attack of fevered remorse. Her long golden hair waved in untended masses over the pillow, like the flames of purgatory; her eyes glared like wheels of fire. Unconscious of what was passing round her, and filled only with her distempered fancies, she cried piteously: 'Drive away those horrid beasts! don't let them come near me! And why do you let those pale-faced creatures pursue me with their hollow glances? If I did deny them food, I cannot help it now! Oh! keep those horrid swine off me! If I did give them the portion of the poor, it is no reason you should let them defile me and trample on me!'

Nothburga was melted with compassion, and her glance of sympathy seemed to chase away the horrid vision. Come to herself, and calm again, Otillia recognized her and begged her pardon, which we may well believe she readily accorded; and shortly after, having reconciled herself to God with true compunction, she fell asleep in peace.[1]

Henry proposed to Nothburga to come and resume

---

[1] I have throughout the story reconciled, as well as I could, the various versions of every episode in which local tradition indulges. One favourite account of Ottilia's end, however, is so different from the one I have selected above, that I cannot forbear giving it also. It represents Ottilia rushing in despair from her bed and wallowing in the enclosure of the pigs, whence, with all Henry's care, she could not be withdrawn alive. All the strength of his retainers was powerless to restrain the beasts' fury, and she was devoured, without leaving a trace behind; only that now and then, on stormy nights, when the pigs are grunting over their evening meal, some memory of their strange repast seems to possess them, and the wail of Ottilia is heard resounding hopelessly through the valley.

her old post in the castle, and moreover to add to it that of superintending the nurture of his only boy. Nothburga gladly accepted his offer, but, in her strict integrity, insisted on accepting no remission from the three years' service under which she had bound herself to Valentine. This concluded, she was received back with open arms at Castle Rottenburg, whither she took with her one of Valentine's daughters to instruct in household duties, that she might be meet to succeed her when her time should come.

Days of peace on earth are not for the saints. Her fight was fought out. The privations she had undergone in sparing her food for the poor, and her subsequent exposure in the field, brought on an illness, under which she shortly after sank. In conformity with her express desire, her body was laid on a bier, to which two young oxen were yoked, and left to follow their own course. The willing beasts tramped straight away over hill and dale and water-course till they came to the village of Eben, then consisting of but a couple of huts of the poor tillers of the soil, and Valentine's homestead; now, a thriving village, its two inns crowded every holiday with peasants, who make their excursions coincide with a visit of devotion to the peasant maiden's shrine. A small field-chapel of St. Ruprecht was then the only place of devotion, but here next morning the body of the holy maiden was found carefully laid at the foot of the altar, and here it was reverently buried, and for centuries it has been honoured by all the country round.[1] In 1434 the Emperor Maximilian, and Christopher, Prince-Bishop of Brixen, built a church over the spot, of which the ancient chapel served as the quire. In 1718 Gaspar Ignatius, Count of Künigl, the then Prince-Bishop, had the remains exhumed, and carried them with pomp to the neighbouring town of Schwatz, where they were left while the church was restored, and an open sarcophagus prepared for them to remain exposed for the veneration of the faithful, which was completed in 1738. In 1838 a centenary festival was observed with great rejoicing, and on March 27, 1862, the cycle of Nothburga's honour was completed in her solemn canonization at Rome.

The lords of Rottenburg had had possession of this territory, and had been the most powerful family of Tirol, ever since the eighth century; one branch extending its sway over the valleys surrounding the Inn, and another branch commanding the country bordering the Etsch; Leuchtenburg and Fleims being the chief fortress-seats of these latter. Their vast power greatly harassed the rulers of Tirol. In every conflict between the native or Austrian princes and the Dukes of Bavaria their influence would always turn the scale, and they often seem to have exercised it simply to show their power. Their family pride grew so high, that it became

---

[1] Grimm has collected (*Deutche Sagen*, Nos. 349 and 350) other versions of the tradition of oxen deciding the sites of shrines which, like the story of the steeple, meets us everywhere. A similar one concerning a camel is given in Stöber's *Legends of Alsace*.

a proverb among the people. It was observed that just during the period of the holy Nothburga's sojourn in the castle the halo of her humble spirit seemed to exercise a charm over their ruling passion. That was no sooner brought to a close than it once more burst forth, and with intenser energy, and by the end of a century more so blinded them that they ventured on an attempt to seize the supreme power over the land. Friedrich *mit der leeren Tasche* was not a prince to lose his rights without a worthy struggle; and then ensued one which was a noteworthy instance of the protection which royalty often afforded to the poor against the oppressions of a selfish aristocracy in the Middle Ages. Friedrich was the idol of the people: in his youth his hardy temperament had made him the companion not only of the mountain huntsman, but even of the mountain hewer of wood. Called to rule over the country, he always stood out manfully for the liberties of the peasant and the burghers of the little struggling communities of Tirol. The lords and knights who found their power thereby restricted were glad to follow the standard of Henry VI., Count of Rottenburg, in his rebellions. Forgetting all patriotism in his struggle for power, Henry called to his aid the Duke of Bavaria, who readily answered his appeal, reckoning that as soon as, by aiding Henry, he had driven Friedrich out, he would shortly after be able to secure the prize for himself.

The Bavarian troops, ever rough and lawless, now

began laying waste the country in ruthless fashion. A Bavarian bishop, moved to compassion by the sufferings of the poor people, though not of his own flock, pleaded so earnestly with the Duke, that he made peace with Friedrich, who was able to inflict due chastisement on Henry, for, powerful as he was, he was no match for him as a leader. He fell prisoner into Friedrich's hands, who magnanimously gave him his liberty; but, according to the laws of the time, his lands and fiefs were forfeit. Though the spirit of the high-minded noble was unbroken, the darling aim of his race which had devolved upon him for execution was defeated; his occupation gone, and his hopes quenched, he wandered about, the last of his race, not caring even to establish himself in any of the fiefs which he held under the Duke of Bavaria, and which consequently yet remained to him.

The history of Henry VI. of Rottenburg has a peculiarly gloomy and fantastic character. Ambitious to a fault, it was one cause of his ill success that he exercised himself in the nobler pursuits of life rather than in the career of arms. Letters of his which are still preserved show that he owed the ascendancy he exercised over his neighbours quite as much to his strength of character and grasp of mind as to his title and riches. No complaint is brought against him in chronicles of the time of niggardliness towards the Church, or of want of uprightness or patience as a judge; he is spoken of as if he had learned to make himself respected as well as feared. But he lived apart in a lofty sphere of his own, seldom mixing in social intercourse, while his refined tastes prevented his becoming an adept in the art of war. Friedrich, on the other hand, who was a hero in the field by his bravery, was also the favourite of the people through his frank and ready-spoken sympathy. Henry had perhaps, on the whole, the finer—certainly the more cultivated—character, but Friedrich was more the man of the time; and it was this doom of succumbing to one to whom he felt himself superior which pressed most heavily on the last of the Rottenburgers. What became of him was never known; consequently many wild stories became current to account for his end : that he never laid his proud head low at the call of death, but yet wanders on round the precincts where he once ruled; that his untamable ambition made him a prey to the Power of Evil, who carried him off, body and soul, to the reward of the proud; that, shunning all sympathy and refusing all assistance, he died, untended and unknown, in a spot far from the habitations of men. It would appear most probable, however, that his death, like his life, was a contrast with the habits of his age : it is thought that, unable to bear his humiliation, he fell by his own hand within a twelvemonth of his defeat.

The deliverance from this powerful vassal, and the falling in of his domains, tended greatly to strengthen and consolidate Friederich's rule over Tirol, and ultimately to render the government of the country more stable, and more beneficial to the people.

Not long after Henry VI.'s disappearance a mysterious fire broke out in the old castle on two separate occasions, laying the greater part of it in ruins. But on each occasion it was noticed that the devouring element, at the height of its fury, spared the litle room which was honoured as that in which the holy Nothburga had dwelt.

A gentler story about this neighbourhood is of a boy tending sheep upon the neighbouring height, who found among some ruins a beautiful bird's-nest. What was his surprise, on examining his treasure, to find it full of broken shells which the fledglings had cast off and left behind them, but shells of a most singular kind. Still greater was his astonishment when, on showing them at home, his parents told him they were no shells, but pieces of precious ore. The affair caused the peasants to search in the neighbourhood, and led to the discovery of one of those veins of metal the working of which brought so great prosperity to Tirol in the fifteenth century, and which are not yet extinct. Their discovery was always by accident, and often by occasion of some curious incident, while the fact that such finds were to be hit upon acted as a strong stimulant to the imagination of a romantic and wonder-loving people, giving belief to all sorts of fables to tell how the treasure was originally deposited, and how subsequently it was preserved and guarded.

CHAPTER III.

NORTH TIROL—UNTERINNTHAL (RIGHT INN-BANK).
THE ZILLERTHAL.

'I may venture to say that among the nations of Europe, and I have more or less seen them all, I do not know any one in which there is so large a measure of real piety as among the Tyroleans. . . . I do not recollect to have once heard in the country an expression savouring of scepticism.'—INGLIS.

THE Zillerthal claims to bear the palm over all the Valleys of Tirol for natural beauty—a claim against which the other valleys may, I think, find something to say.

There is an organised service of carriages (the road is only good for an *einspanner*—one-horse vehicle) into the Zillerthal, at both Brixlegg and Jenbach, taking between four and five hours to reach Zell, an hour and a-half more to Mayrhofen. Its greatest ornaments are the castles of Kropfsberg, Lichtwer, and Matzen; the Reiterkogel and the Gerlos mountains, forming the present boundary against Salzburg; and the Ziller, with its rapid current which gave it its name (from *celer*),[1]

---

[1] It is perhaps to be reckoned among the tokens of Etruscan residence among the Rhætian Alps, for Mr. Isaac Taylor finds that the word belongs to their language. (*Etruscan Researches*, pp. 333, 380.)

its tributary streams might very well have received the same appellation, for their *celerity* is often so impetuous that great damage is done to the inhabitants of the neighbourhood.

Before starting for the Zillerthal I may mention two castles which may also be seen from Jenbach, though like it they belong in strictness to the chapter on the Left Inn-bank. One is Thurnegg by name, which was restored as a hunting-seat by Archduke Ferdinand; and at the instance of his second wife, the pious Anna Katharina of Mantua, he added a chapel, in order that his hunting-parties might always have the opportunity of hearing Mass before setting out for their sport.

Another is Tratzberg, which derived its name from its defiant character. It is situated within an easy walk of Jenbach. Permission to visit it is readily given, for it counts as a show-place. It may be taken on the way to S. Georgenberg and Viecht, but it occupies too much time, and quite merits the separate excursion by its collections and its views. Frederick sold it in 1470 to Christian Tänzel, a rich mining proprietor of the neighbourhood, who purchased with it the right to bear the title of Knight of Tratzberg. No expense was spared in its decoration, and its paintings and marbles made it the wonder of the country round. In 1573 it passed into the hands of the Fuggers, and at the present day belongs to Count Enzenberg, who makes it an occasional residence. A story is told of it which is in striking contrast to that mentioned of Thurnegg. One of the knights of the castle in ancient time had a reputation for caring more for the pleasures of the chase than for the observances of religion. Though he could get up at an early hour enough at the call of his *Jäger's* horn, the chapel bell vainly wooed him to Mass.

In vain morning by morning his guardian angel directed the sacred sound upon his ear; the knight only rolled himself up more warmly in the coverlet, and said, 'No need to stir yet, the dogs are not brought round till five o'clock.'

'Ding—dong—dang! Come—to—Mass! Ding—dong—dang!' sang the bells.

'No, I can't,' yawned the knight, and covered his ear with the bed-clothes.

The bell was silent, and the knight knew that the pious people who had to work hard all day for their living, and yet spared half an hour to ask God's blessing on their labours, were gone into the chapel.

He fancied he saw the venerable old chaplain bowing before the altar, and smiting his breast; he saw the faithful rise from their knees while the glad tidings of the Gospel were announced, and they proclaimed their faith in them in the Creed; he heard them fall on their knees again while the sacred elements were offered on the altar and the solemn words of the consecration pronounced; he saw little Johann, the farrier's son, bow his head reverently on the steps, and then sound the threefold bell which told of the most solemn moment of the sacred mysteries; and the chapel bell took up the note, and announced the joyful news to those whom illness or necessity forced to remain away.

Then hark! what was that? The rocks under the foundation of the castle rattled together, and all the stones of its massive walls chattered like the teeth of an old woman stricken with fear. The three hundred and sixty-five windows of the edifice rattled in their casements, but above them all sounded the piercing sound of the knight's cry of anguish. The affrighted people rushed into the knight's chamber; and what was their horror when, still sunk in the soft couch where he was wont to take his ease, there he lay dead, while his throat displayed the print of three black and burning claws. The lesson they drew was that the knight, having received from his guardian angel the impulse to repair his sloth by at least *then* rising to pay the homage which the bell enjoined, had rejected even this last good counsel, thereby filling up the measure of his faults. For years after marks were shown upon the wall as having been sprinkled by his blood!

The first little town that reckons in the Zillerthal is Strass, a very unpretending place, and then Schlitters.

At Schlitters they have a story of a butcher who, going to Strass to buy an ox, had scarcely crossed the Zill and got a little way from home, than he saw lying by the way-side a heap of the finest wheat. Not liking to appropriate property which might have a legitimate owner, he contented himself with putting a few grains in his pocket, and a few into his sack, as a specimen.

As he went by the way his pockets and his sack began to get heavier and heavier, till it seemed as if the weight would burst them through. Astonished at the circumstance he put in his hand, and found them all full of shining gold. As soon as he had recovered his composure, he set off at the top of his speed, and, heeding neither hill or dale, regained the spot where he had first seen the wheat. But it was no more to be seen. If he had had faith to commend himself to God on his first surprise, say the peasants, and made the holy sign of redemption, the whole treasure would have been his.

There is another tradition at Schlitters of a more peculiar character. It is confidently affirmed that the village once boasted two churches, though but a very small one would supply the needs of the inhabitants. Hormayr has sifted the matter to the bottom, and explains it in this way. There lived in the neighbourhood two knights, one belonging to the Rottenburger, and the other to the Freundsberger family. Now the latter had a position of greater importance, but the former possessed a full share of family haughtiness, and would not yield precedence to any one. In order not to be placed on a footing of inferiority, or even of equality, with his rival, he built a second church, which he might attend without being brought into contact with him. No expense was spared, and the church was solidly built enough; but no blessing seemed to come on the edifice so built, no pains could ever keep it in repair, and at last, after crumbling into ruin, every stone of it disappeared.

Kropfsberg is a fine ruin, belonging to Count Enzenberg, seen a little above Strass, on a commanding height between the high road and the Inn. It is endeared to the memory of the Tiroleans by having been the spot where, on St. Michael's Day, 1416, their favourite Friedrich *mit der leeren Tasche* was reconciled with his brother Ernst *der Eiserne*, who, after the Council of Constance had pronounced its ban on Frederick, had thought to possess himself of his dominions.

The largest town of the Zillerthal is Fügen, a short distance below Schlitters, and the people are so proud of it, that they have a saying ever in their mouths, 'There is but one Vienna and one Fügen in the world!' It doubtless owes its comparative liveliness and prosperity to its château being kept up and often inhabited by its owners (the Countess of Dönhof and her family). This is also a great ornament to the place, having been originally built in the fifteenth century by the lords of Fieger, though unhappily the period of its rebuilding (1733) was not one very propitious to its style. The sculpture in the church by the native artist, Nissl, is much more meritorious. The church of Ried, a little further along the valley, is adorned with several very creditable pictures by native artists. It is the native place of one of the bravest of the defenders of throne and country, so celebrated in local annals of the early part of the century, Sebastian Riedl. He was only thirty-nine at his death in 1821. Once, on an occasion of his fulfilling a mission to General

Blucher, he received from him a present of a hussar's jacket, which he wore at the battle of Katzbach, and it is still shown with pride by his compatriots.

The Zillerthal was the only part of Tirol where Lutheranism ever obtained any hold over the people. The population was very thin and scattered, consequently they were out of the way of the regular means of instruction in their own faith; and it often happened, when their dwellings and lands were devastated by inundations, that they were driven to seek a livelihood by carrying gloves, bags, and other articles made of chamois leather, also of the horns of goats and cattle, into the neighbouring states of Germany. Hence they often came back imbued with the new doctrines, and bringing books with them, which may have spread them further. This went on, though without attracting much attention, till the year 1830, when they demanded permission to erect a church of their own. The *Stände* of Tirol were unanimous, however, to resist any infringement of the unity of belief which had so long been preserved in the country. The Emperor confirmed their decision, and gave the schismatics the option of being reconciled with the Church, or of following their opinions in other localities of the empire where Lutheran communities already existed. A considerable number chose the latter alternative, and peace was restored to the Zillerthal. Every facility was given them by the government for making the move advantageously, and the inhabitants, who had been long provoked by the scorn and ridicule with which the exiles had treated their time-honoured observances, held a rejoicing at the deliverance.

At the farther end of the valley is Zell, which though smaller in population than Fügen, has come to be considered its chief town. Its principal inn, for there are several—*zum Post*—if I recollect right, claims to be not merely a *Gasthaus*, but a *Gasthof*. The *Brauhaus*, however, with less pretension, is a charming resort of the old-fashioned style, under the paternal management of Franz Eigner, whose daughters sing their local melodies with great zest and taste. The church, dedicated to St. Vitus, is modern, having been built in 1771-82; but its slender green steeple is not inelegant. It contains some meritorious frescoes by Zeiler. The town contains some most picturesque buildings, as the Presbytery, grandiloquently styled the *Dechanthof*, one or two educational establishments, several well-to-do private houses, and the town-hall, once a flourishing brewery, which failed—I can hardly guess how, for the chief industry of the place is supplying the neighbourhood with beer.

A mile beyond Zell is Hainzenberg, where the process of gold-washing on a small scale may be studied, said to be carried on by the owner, the Bishop of Brizen, on a sort of ultra-co-operative principle, as a means of support to the people of the place, without profit to himself. There is also a rather fine waterfall in the neighbourhood, and an inn where luncheon may be had. The most interesting circumstance, perhaps, in connexion with Zell is the *Kirchweih-fest*, which is very celebrated in all the country round. I was not fortunate enough to be in the neighbourhood at the right time of year to witness it. On the other side of the Hainzenberg, where the mountain climber can take his start for the Gerlozalp, is a little sanctuary called *Mariä-rastkapelle*, and behind it runs a sparkling brook. Of the chapel the following singular account is given:—In olden time there stood near the stream a patriarchal oak sacred to Hulda;[1] after the introduction of Christianity the tree was hewn down, and as they felled it they heard Hulda cry out from within. The people wanted to build up a chapel on the spot in honour of the Blessed Virgin, and began to collect the materials. No sooner had the labourers left their work, however, than there appeared an army of ravens, who, setting themselves vigorously to the task, carried every stone and every balk of wood to a neighbouring spot. This happened day after day, till at last the people took it as a sign that the soil profaned by the worship of Hulda was not pleasing to heaven, and so they raised their chapel on the place pointed out by the ravens, where it now stands.

After Mayrhof, the next village (with three inns), in the neighbourhood of which garnets are found and

[1] 'Hulda was supposed to delight in the neighbourhood of lakes and streams; her glittering mansion was under the blue waters, and at the hour of mid-day she might be seen in the form of a beautiful woman bathing and then disappearing.'—Wolf, *Deutsche Götterlehre*. See also Grimm, *Deutsche Mythologie*, pp. 164-8.

mills for working them abound, the Zillerthal spreads out into numerous branches of great picturesqueness, but adapted only to the hardy pedestrian, as the Floitenthal, the Sondergrundthal, the Hundskehlthal (Dog's-throat valley), the Stillupethal, with its Teufelsteg, a bridge spanning a giddy ravine, and its dashing series of waterfalls. The whole closed in by the Zemmer range and its glaciers, the boundary against South-Tirol, said to contain some of the finest scenery and best hunting-grounds in the country. It has been also called the 'el Dorado' of the botanist and the mineralogist. The most important of these by-valleys is the Duxerthal, by non-Tiroleans generally written Tuxerthal, a very high-lying tract of country, and consequently one of the coldest and wildest districts of Tirol. Nevertheless, its enclosed and secluded retreat retains a saying perhaps many thousand years old, that once it was a bright and fertile spot yielding the richest pastures, and that then the population grew so wanton in their abundance that they wasted their substance. Then there came upon them from above an icy blast, before which their children and their young cattle sank down and died; and the herbage was, as it were, bound up, and the earth was hardened, so that it only brought forth scarce and stunted herbs, and the mountain which bounded their pleasant valley itself turned to ice, and is called to this day *die gefrorene Wand*, the frozen wall. The scattered population of this remote valley numbered so few souls, that they depended on neighbouring villages for their ecclesiastical care, and during winter when shut in by the snow within their natural fastnesses, were cut off from all spiritual ministration, so that the bodies of those who died were preserved in a large chest, of which the remains are yet shown, until the spring made their removal to Mattrey possible. In the middle of the seventeenth century they numbered 645 souls, and have now increased to about 1,400; about the year 1686 they built a church of their own, which is now served by two or three priests. For the first couple of miles the valley sides are so steep, that the only level ground between them is the bed of an oft-times torrential stream, but yet they are covered almost to the very top with a certain kind of verdure; further on it widens out into the district of Hinterdux, which is a comparatively pleasant cheerful spot, with some of the small cattle (which are reared here as better adapted to the gradients on which they have to find their food,) browsing about, and sundry goats and sheep, quite at home on the steeps. But scarce a tree or shrub is to be seen—just a few firs, and here and there a solitary mountain pine; and in the coldest season the greatest suffering is experienced from want of wood to burn. The only resource is grubbing up the roots remaining from that earlier happier time, which but for this proof might have been deemed fabulous.

The hardships which the inhabitants of this valley cheerfully undergo ought to serve as a lesson of diligence indeed. The whole grass-bearing soil is divided among them. The more prosperous have a cow or more of their own, by the produce of which they live; others take in cows from Innsbruck and Hall to graze. The butter they make becomes an article of merchandise, the transport of which over the mountain paths provides a hard and precarious livelihood for a yet poorer class; the pay is about a halfpenny per lb. per day, and to make the wage eke out a man will carry a hundred and a woman fifty to seventy pounds through all weathers and over dangerous paths, sleeping by night on the hard ground, the chance of a bundle of hay in winter being a luxury; and one of their snow-covered peaks is with a certain irony named the Federbett. They make some six or seven cwt. of cheese in the year, but this is kept entirely for home consumption.

The care of these cattle involves a labour which only the strongest constitution could stand—a continual climbing of mountains in the cold, often in the dark, during great part of the year allowing scarcely four or five hours for sleep. Nor is this their only industry. They contrive also to grow barley and flax; this never ripens, yet they make from it a kind of yarn, which finds a ready sale in Innsbruck; they weave from it too a coarse linen, which helps to clothe them, together with the home-spun wool of their sheep. Also, by an incredible exercise of patience, they manage to heap up and support a sufficient quantity of earth round the rough and stony soil of their valley to set potatoes, carrots, and other roots. Notwithstanding all these hardships, they are generally a healthy race, remarkable for their endurance, frugality, and love of home. Neither does their hard life make them neglect the improvement of the mind; nowhere are schools more regularly attended, although the little children have many of them an hour or two's walk through the snow. The church is equally frequented; so that if the great cold be sent, as the legend teaches, as a chastisement,[1] the people seem to have had grace given them to turn it to good account.

The Zemgrund, Zamsergrund, and the Schwarzensteingrund, are other pedestrian excursions much recommended from Mayrhof, but all equally require the aid of local guides, and have less to repay toil than those already described.

Travellers who merely pass through Tirol by rail may catch a sight of the mountains which hem in the Duxerthal, just after passing the station of Steinach, on their left hand, when facing the south.

[1] One version of the legend says, the Frozen Wall was formed out of the quantities of butter the people had wasted.

CHAPTER IV.

NORTH TIROL—UNTERINNTHAL (RIGHT INN-BANK).

(ZILLERTHAL CUSTOMS.—THE WILDSCHÖNAU.)

Deep secret springs lie buried in man's heart,
Which Nature's varied aspect works at will ;
Whether bright hues or shadows she impart,
Or fragrant odours from her breath distil,
Or the clear air with sounds melodious fill ;
She speaks a language with instruction fraught,
And Art from Nature steals her mimic skill,
Whose birds, whose rills, whose sighing winds first taught
That sound can charm the soul, and rouse each noble thought.
                                                                        LADY CHARLOTTE BURY.

WE had parted from the Zillerthal, and had once more taken our places in the railway carriage at Jenbach for a short stage to reach Kundl,[1] as a base of operations for visiting the Wildschönau, as well as the country on the other side of the Inn. The entry was effected with the haste usual at small stations, where the advent of a traveller, much more of a party of tourists, is an exceptional event. The adjustment of our bags and rugs was greatly facilitated by the assistance of the only occupant of the compartment into which we were

---

[1] This excursion was made on occasion of a different journey from that mentioned in Chapter i. Of course, if taken on the way from Kufstein to Innsbruck, you would take the Wildschönau *before* the Zillerthal.

thrust; and when we had settled down and expressed our thanks for his urbanity, I observed that he eyed us with an amused but not unpleasant scrutiny. At last his curiosity overcame his reticence. 'I have frequent occasion to travel this way to Munich and Vienna,' he said, 'and I do not remember ever to have fallen in with any strangers starting from Jenbach.'

The conversation so opened soon revealed that our new friend, though spending most of his time in the Bavarian and Austrian capitals, nevertheless retained all a mountaineer's fondness for the Tirolese land, which had given him birth some seventy years before. He was greatly interested in our exploration of the Zillerthal, but much annoyed that we were leaving instead of entering it; had it been the other way, he said, he would have afforded us an acquaintance with local customs such as, he was sure, no other part of Europe could outvie. I assured him I had been disappointed at not coming across them during our brief visit, but fully hoped on some future occasion to have better success. He warmly recommended me not to omit the attempt, and for my encouragement cited a local adage testifying to the attractions of the valley—

<blockquote>
Wer da kommt in's Zillerthatl
Der kommt gewiss zum Zweitenmal.[1]
</blockquote>

He was interesting us much in his vividly-coloured sketches of peasant life, when the train came to a

---

[1] Whoever comes into the Zillerthal is sure to visit it a second time.

stand; the guard shouted 'Kundl,' and we were forced to part. He gave us an address in Munich, however, where we were afterwards fortunate enough to find him; and he then gave me some precious particulars, which I was not slow to garner.

He seemed to know the people well, having lived much among them in his younger days, and claimed for them—perhaps with some little partiality—the character of being industrious, temperate, moral, and straightforward, even above the other dwellers in Tirol; and no less, of being physically the finest race. Their pure bracing mountain air, the severe struggle which nature wages with them in their cultivation of the fruits of the soil, and the hardy athletic pursuits with which they vary their round of agricultural labour, tend to maintain and ever invigorate this original stock of healthfulness. Their athletic games are indeed an institution to which they owe much, and which they keep up with a devotion only second to that with which they cultivate their religious observances. Every national and social festival is celebrated with these games. The favourite is the *scheibenschiessen*, or shooting at a mark, for accuracy in which they are celebrated in common with the inhabitants of all other districts of the country, but are beaten by none; their *stutze* (short-barrelled rifle) they regard more in the light of a friend and companion than a weapon, and dignify it with the household name of the bread-winner. Wrestling is another favourite sport; to be the champion wrestler of the hamlet is a distinction which no inhabitant of the Zillerthal would barter for gold. The best ' *Haggler*,' ' *Mairraffer*,' and ' *Roblar* '— three denominations of wrestlers—are regarded somewhat in the light of a superior order of persons, and command universal respect. In wilder times, it is true, this ran into abuse; and some who had attained excellence in an art so dangerous when misapplied betook themselves to a life of violence and freebooting; but this has entirely passed away now, and anything like a highway robbery is unheard of. The most chivalrous rules guard the decorum of the game, which every bystander feels it a point of honour to maintain; the use even of the *stossring*, a stout metal ring for the little finger, by which a telling and sometimes disfiguring blow may be given by a dexterous hand, is discouraged. It is still worn, however, and prized more than as a mere ornament—as a challenge of the wearer's power to wield it if he choose, or if provoked to show his prowess. Running in races—which, I know not why, they call *springen*—obtains favour at some seasons of the year. At bowls and skittles, too, they are famous hands; and in their passion for the games have originated a number of fantastic stories of how the fairies and wild men of the woods indulge in them too. Many a herdsman, on his long and solitary watch upon the distant heights, gives to the noises of nature which he has heard, but could not account for, an origin which lives in the imagination of those to whom he recounts it on his return home; and his fancies are recorded as actual events. But that the spirits play at skittles, and with gold and silver balls, is further confirmed by peasants who have lost their way in mists and snow-storms, and whose troubled dreams have made pleasant stories. One of these, travelling with his pedlar's pack, sought refuge from the night air in the ruined castle of Starkenberg, the proud stronghold of a feudal family, second only in importance to the Rottenburgers, and equally brought low by Friedrich *mit der leeren Tasche*. The pedlar was a bold wrestler, and felt no fear of the airy haunters of ruined castles. He made a pillow of his pack, and laid him down to sleep as cosily as if at home, in the long dank grass; nevertheless, when the clock of the distant village church—to whose striking he had been listening hour by hour with joy, as an earnest that by the morning light he would know how to follow its guiding to the inhabited locality it denoted—sang out the hour of midnight, twelve figures in ancient armour stalked into the hall, and set themselves to play at bowls, for which they were served with skulls. The pedlar was a famous player, and nothing daunted, took up a skull, and set himself to play against them, and beat them all; then there was a shout of joy, such as mortal ears had never heard, and the twelve spirits declared they were released. Scarcely had they disappeared, when ten more spirits, whom the pedlar concluded like the last to be retainers of the mighty Starkenberger of old, entered by different doors, which they carefully locked behind them, and then bringing our hero the keys, begged him to open the doors each with the right one. The pedlar was a shrewd fellow; and though doors, keys, and spirits were each alike of their kind, his observation had been so accurate that he opened each with the right key without hesitation, whereupon the ten spirits declared themselves released too. Then came in the Evil One, furious with the pedlar, who was setting free all his captives, and swore he would have him in their stead. But the pedlar demanded fair play, and offered to stake his freedom on a game with his Arch-Impiety. The pedlar won, and the demon withdrew in ignominy; but the released spirits came round their deliverer, and loaded him with as much gold and valuable spoil as he could carry.

This story seemed to me to belong to a class not unfrequently met with, but yet differing from the ordinary run of legends on this subject, inasmuch as the spirits, who were generally believed to be bound to earth in penance, were released by no act of Christian virtue, and without any appeal to faith; and I could not help asking my old friend if he did not think this very active clever pedlar might have been one of those who according to his own version had indulged in free-booting tendencies, and that having with a true Zillerthaler's tendencies pined to return to his native valley, he had invented the tale to account for his accession of fortune, and the nature of his possessions.

I think my friend was a little piqued at my unmasking his hero, but he allowed it was not an improbable solution for the origin of some similar tales.

Prizes, he went on to tell me, are often set up for excellence in these games, which are cherished as marks of honour, without any reference to their intrinsic value. And so jealously is every distinction guarded, that a youth may not wear a feather or the sprig of rosemary, bestowed by a beloved hand, in his jaunty hat, unless he is capable of proving his right to it by his pluck and muscular development.

Dancing is another favourite recreation, and is pursued with a zest which makes it a healthful and useful exercise too. The *Schnodahüpfl* and the *Hosennagler* are as dear to the Zillerthaler as the *Bolera* to the Andalusian or the *Jota* to the Aragonese; like the Spanish *Seguidillas*, too, the Zillerthalers accompany their dance with sprightly songs, which are often directed to inciting each other not to flag.

Another amusement, in which they have a certain similarity with Spaniards, is cow-fighting. But it is not a mere sport, and cruelty is as much avoided as possible, for the beasts are made to fight only with each other, and only their natural weapons—each other's horns—are brought against them. The victorious cow is not only the glory and darling of her owner, who loads her with garlands and caresses; but the fight serves to ascertain the hardy capacity of the animals as leaders of the herd, an office which is no sinecure, when they have to make their way to and from steep pastures difficult of access.[1] Ram and goat fights are also held in the same way, and with the same object.

The chief occasions for exercising these pastimes are the village festivals, the *Kirchtag*, or anniversary of the Church consecration, the Carnival season, weddings and baptisms, and the opening of the season for the *Scheibenchiessen*; also the days of pilgrimages to various popular shrines; and the *Primizen* and *Sekundizen*—the first Mass of their pastors, and its fiftieth anniversary—general festivals all over Tirol.

A season of great enjoyment is the Carnival, which with them begins at the Epiphany. Their great delight then is to go out in the dusk of evening, when work is over, disguised in various fantastic dresses, and making their way round from house to house, set the inmates guessing who they can be. As they are very clever in arranging all the accessories of their assumed character, changing their voice and mien, each visit is the occasion of the most laughable mistakes. In the towns, the Carnival procession is generally got up with no little taste and artistic skill. The arch-buffoon goes on ahead, a loud and merry jingle of bells announcing his advent at every movement of the horse he bestrides, collects the people out of every house. Then follow,

---

[1] In the Vintschgau (see *infra*) the leading cow has the title of *Proglerin*, from the dialectic word *proglen*, to carry one's head high. She wears also the most resounding bell.

also mounted, a train of maskers, Turks, soldiers, gipsies, pirates; and if there happen to be among them anyone representing a judge or authority of any sort, he is always placed at the head of the tribe. In the evening, their perambulations over, they assemble in the inn, where the acknowledged wag of the locality reads a humorous diatribe, which touches on all the follies and events, that can be anyhow made to wear a ridiculous aspect, of the past year.

Christmas—here called *Christnacht* as well as *Weihnacht*—is observed (as all over the country, but especially here) by dispensing the *Kloubabrod*, a kind of dough cake, stuffed with sliced pears, almonds, nuts, and preserved fruits. The making of this is a particular item in the education of a Zillerthaler maiden, who has a special interest in it, inasmuch as the one she prepares for the household must have the first cut in it made by her betrothed, who at the same time gives her some little token of his affection in return. Speaking of Christmas customs reminded my informant of an olden custom in Brixen, that the Bishop should make presents of fish to his retainers. This fish was brought from the Garda-see, and the Graf of Tirol and the Prince-bishop of Trent were wont to let it pass toll-free through their dominions. A curious letter is extant, written by Bishop Rötel, '*an sambstag nach Stæ. Barbaræ*, 1444,' courteously enforcing this privilege.

The *Sternsingen* is a favourite way of keeping the Epiphany in many parts of the country. Three youths, one of them with his face blackened, and all dressed to represent the three kings, go about singing from homestead to homestead; and in some places there is a Herod ready to greet them from the window with riming answers to their verses, of which the following is a specimen: it is the address of the first king—

> König Kaspar bin ich gennant
> Komm daher aus Morgenland
> Komm daher in grossen Eil
> Vierzehn Tag, fünftausend Meil.
> Melchores tritt du herein.[1]

Melchior, thus appealed to, stands forward and sings his lay; and then Balthazar; and then the three join in a chorus, in which certain hints are given that as they come from so far some refreshment would be acceptable; upon which the friendly peasant-wife calls them in, and regales them with cakes she has prepared ready for the purpose, and sends them on their mountain-way rejoicing. Possibly some such custom may have given rise to the institution of our 'Twelfth-cake.' In the Œtzthal they go about with the greeting, 'Gelobt sei Jesus Christus zur Gömacht.'[2] Another Tirolean custom connected with Epiphany was the blessing of the stalls of the cattle on the eve, in memory of the stable in which the Wise Men found the Holy Family.

---

[1] 'Kaspar my name: from the East I came: I came thence with great speed: five thousand miles in fourteen days: Melchior, step in.' Zingerle gives a version of the whole set of rimes.

[2] See *Sitten Bräuche u. Meinungen des Tiroler Volks*, p. 81.

Their wedding fêtes seem to be among the most curious of all their customs. My friend gave me a detailed account of one, between two families of the better class of peasants, which he had attended some years back; and he believed they were little changed since. It is regarded as an occasion of great importance; and as soon as the banns had been asked in church, the bridegroom went round with a chosen friend styled a *Hochzeitsbitter*, to invite friends and relations to the marriage. The night before the wedding (for which throughout Tirol a Thursday is chosen, except in the Iselthal, where a preference for Monday prevails), there was a great dance at the house of the bride, who from the moment the banns have been asked is popularly called the *Kanzel-Braut*. 'Rather, I should say,' he pursued, 'it was in the barn; for though a large cottage, there was no room that would contain the numbers of merry couples who flocked in, and even the barn was so crowded, that the dancers could but make their way with difficulty, and were continually tumbling over one another; but it was a merry night, for all were in their local costume, and the pine-wood torches shed a strange and festive glare over them. The next morning all were assembled betimes. It was a bitterly cold day, but the snow-storm was eagerly hailed, as it is reckoned a token that the newly-wedded pair will be rich; we met first at the bride's house for what they called the *Morgensuppe*, a rough sort of hearty breakfast of roast meat, white bread, and sausages; and while the elder guests were discussing it, many were hard at work again dancing, and the young girls of the village were dressing up the bride—one of the adornments *de rigueur* being a knot of streamers of scarlet leather trimmed with gold lace, and blue arm-bands and hat-ribbons; these streamers are thought by the simple people to be a cure for goitres, and are frequently bound round them with that idea. At ten o'clock the first church bell rang, when all the guests hastily assembled round the table, and drank the health of the happy pair in a bowl from which they had first drank. Then they ranged themselves into a procession, and marched towards the church, the musicians leading the way. The nearest friends of the bridal pair were styled "train-bearers," and formed a sort of guard of honour round the bride, walking bare-headed, their hats, tastily wreathed with flowers, in their hands. The priest of the village walked by the bride on one side, her parents on the other. She wore a wreath of rosemary—a plant greatly prized here, as among the people of Spain and Italy, and considered typical of the Blessed Virgin's purity—in her hair; her holiday dress was confined by a girdle, and she held her rosary in her hand. The bridegroom was almost as showily dressed, and wore a crown of silver wire; beside him walked another priest, and behind them came the host of the village inn, a worthy who holds a kind of patriarchal position in our villages. He is always one of the most important men of the place, generally owns the largest holding of land,

and drives one or two little trades besides attending to the welfare of his guests. But more than this, he is for the most part a man of upright character and pleasant disposition, and is often called to act as adviser and umpire in rural complications.

'The procession was closed by the friends and neighbours, walking two and two, husband and wife together; and the church bells rang merrily through the valley as it passed along.

'The ceremonial in the church was accompanied with the best music the locality could afford, the best singers from the neighbouring choirs lending their voices. To add to the solemnity of the occasion, lighted tapers were held by the bridal party at the Elevation; and it was amusing to observe how the young people shunned a candle that did not burn brightly, as that is held to be an omen of not getting married within the year. At the close of the function, the priest handed round to them the *Johannissegen,* a cup of spiced wine mixed with water, which he had previously blessed, probably so called in memory of the miracle at the wedding-feast recorded in the Gospel of that Apostle.

'The band then struck up its most jocund air, and full of mirth the gladsome party wended their way to the inn. After a light repast and a short dance, and a blithesome *Trutzlied,* they passed on, according to custom, to the next, and so on to all the inns within a radius of a few miles. This absorbed about three or four hours; and then came the real wedding banquet, which was a very solid and long affair—in fact, I think fresh dishes were being brought in one after another for three or four hours more. Even in this there was a memory of the Gospel narrative, for in token of their joy they keep for the occasion a fatted calf, the whole of which is served up joint by joint, not omitting the head; this was preceded by soup, and followed by a second course of sweet dumplings, with fruit and the inevitable pickled cabbage, which on this day is dignified with the title of *Ehrenkraut*. After this came a pause; and the musicians, who had been playing their loudest hitherto, held in too. The "best man" rose, and went through the formula of asking the guests whether they were content with what had been set before them, which of course was drowned in a tumult of applause. In a form, which serves from generation to generation with slight change, he then went on to remark that the good gifts of meat and drink of which they had partaken came from the hand of God, and called forth the gratitude of the receiver, adding, "Let us thank Him for them, and still more in that He has made us reasonable beings, gifting us with faith, and not brutes or unbelievers. If we turn to Him in this spirit, He will abide with us as with them of Cana in Galilee. Therefore, let all anger and malice and evil speaking be put away from us, who have just been standing before the most holy Sacrament, and let us be united in the bonds of brotherly love, that His

Blood may not have been poured out for us in vain. And to you, dear friends, who have this day been united with the grace-giving benediction of the Church, I commend this union of heart and soul most of all, that the new family thus founded in our midst may help to build up the living edifice of a people praising and serving God, and that you walk in His way, and bring up children to serve Him as our forefathers have ever done." There was a good deal more in the same strain; and this exhortation to holy living, from one of themselves, is just a type of the intimate way in which religion enters into the life of the people. His concluding wish for the well-being of the newly married was followed by a loud "Our Father" and "Hail Mary" from the assembled throng.

'After this came a great number more dishes of edibles, but this time of a lighter kind; among them liver and poultry, but chiefly fruits and sweets; and among these many confections of curious devices, mostly with some symbolical meaning. When these were nearly despatched, wine and brandy were brought out by the host; and by this name you must understand the master of the inn; for, true to the paternal character of which I have already spoken, it is always his business to cater for and preside over bridal banquets. At the same time the guests produced their presents, which go by the name of *Waisat*, and all were set down in a circumstantial catalogue. They are generally meted out with an open hand, and are a great help to the young people in beginning their housekeeping.

'The musicians, who only got hasty snatches of the good things passing round, now began yet livelier strains, and the party broke up that the younger members might give themselves to their favourite pastime, dancing; and well enough they looked, the lads in brilliant red double-breasted waistcoats, their short black leather breeches held up with embroidered belts, and their well-formed high-pointed hats with jaunty brim, going through the intricate evolutions, each beating the time heartily, first on his thighs and then on his feet—*schuhplatteln* they call it—and followed through the mazy figures by his *diandl* (damsel), in daintily fitting satin bodice, and short but ample skirt.

'The older people still lingered over the table, and looked on at the dance, which they follow with great interest; but there is not a great deal of drinking, and it is seldom enough, even in the midst of an occasion for such exceptional good cheer, that any excess is committed. A taste for brandy—the poor brandy of their own manufacture—is however, I confess, a weakness of the Zillerthalers. The necessity for occasionally having recourse to stimulants results from the severity of the climate during part of the year, and the frequently long exposure to the mountain air which their calling requires of them. At the same time, anything like a confirmed drunkard is scarcely known among them. Its manufacture affords to many an occupation; and its use to all, of both sexes, is a national habit. They make it out of barley, juniper, and numbers of other berries (which they wander collecting over all the neighbouring alps), as well as rye, potatoes, and other roots—in fact, almost anything. Every commercial bargain, every operation in the field, every neighbourly discussion, every declaration of affection even, is made under its afflatus. An offer of a glass of the cordial will often make up a long-harboured quarrel, a refusal to share one is taken to be a studied affront; in fact, this *zutrinken*, as they call it, comes into every act and relation of life. In the moderate bounds within which they keep its use, it is undeniably a great boon to them; and many a time it has been the saving of life in the mountains to the shepherd and the milkmaid, the snow-bound labourer or retarded pedlar.'

I was curious to know what customs the other valley had to replace those of the Ziller. My friend informed me they were very similar, only the Ziller-thalers were celebrated for their attachment to and punctual observance of them. He had once attended a wedding in the Grödnerthal which was very similar to the one he had already described, yet had some distinct peculiarities. Though a little out of place, I may as well bring in his account of it here. There, the betrothal is called *der Handschlag* (*lit.* the hand-clasp), and it is always performed on a Saturday. The fathers of the bride and bridegroom and other nearest relations are always present as witnesses; and if the bride does not cry at the projected parting, it is said she will have many tears to shed during her married life. The first time the banns are asked it is not considered 'the thing' for the betrothed to be present, and they usually go to church on that occasion in some neighbouring village; on the second Sunday they are expected to appear in state, the bridegroom wearing his holiday clothes and a nosegay in his hat or on his right breast. The bride always wears the local costume, a broadish brimmed green hat, a scarlet boddice and full black skirt, though this is now only worn on such occasions; on the day of the wedding, to this is added a broad black satin ribbon round her head, and round her waist a leather girdle with a number of useful articles in plated copper hanging from it. On each side are arranged red and green streamers with very great nicety, and no change of fashion is suffered in their position; she is expected to wear a grave mien and modest deportment; this is particularly enjoined. The guests are also expected to don the popular costume; the girls green, the married women black hats. On the way to the church the bridegroom's father and his nearest neighbour came forward, and with many ceremonies asked the bride of her friends, and she went crying coyly with them. After the church ceremony, which concludes as in Zillerthal with the cup of S. Johannessegen, the bridesmaids hand in a basket decked with knots of ribbon, containing offerings for the priests and servers, and a wreath, which is fastened round the priest's arm who leads the bride out of church. The visit to the neighbouring inn follows; but at the wedding feast guests come in in masquerading dresses bringing all manner of comical presents. The dance here lasts till midnight, when the happy pair are led home by their friends to an accompaniment of music, for which they have a special melody. The next day again there are games, and the newly married go in procession with their friends to bear home the trousseau and wedding gifts, among which is always a bed and bedding. On their way back beggars are allowed to bar the way at intervals, who must be bought off with alms. On the Sunday following the bride is expected again to appear at church in the local costume, and in the afternoon all the guests of the wedding day again gather in the inn to present their final offering of good wishes and blessings. Girls who are fond of cats, they say, are sure to marry early; perhaps an evidence that household virtues are appreciated in them by the men; but of men, the contrary is predicated, showing that the other sex is expected to display hardihood in the various mountaineering and other out-door occupations.[1]

Kundl, whither we were bound before being

[1] Its origin may be traced further back than this, perhaps. The cat was held to be the sacred animal of Freia (Schrader, *Germ. Myth.*), and the word *freien*, to woo, to court, is derived from her name. (Nork.)

tempted to make this digression, gives entrance to the Wildschönau according to modern orthography, the *Witschnau*, or *Wiltschnau*, according to local and more correct pronunciation (sometimes corrupted into Mitschnau), as the name is derived from *wiltschen*, to flow, and *au*, water, the particular water in this case being the Kundler-Ache, which here flows into the Inn. It is a little valley improving in beauty as you pursue it eastwards, not more than seven leagues in length, and seldom visited, for its roads are really only fit for pedestrians; hence its secluded inhabitants have acquired a character for being suspicious of strangers, though proverbially hospitable to one another. One of its points of greatest interest is the church of St. Leonhard, described in the last chapter. Overhanging the road leading from it to Kundl, stand the remains of the castle of Niederaich, now converted into a farm stable, and its moat serving as a conduit of water for the cattle. At the time it was built by Ambrose Blank in the sixteenth century, the silver mines then in work made this a most flourishing locality. At that time, too, there stood overlooking the town the Kundlburg, of which still slighter traces remain, the residence of the Kummerspruggers, who, in the various wars, always supported the house of Bavaria. The chief industry of Kundl at present is the construction of the boats which navigate the Inn, and carry the rich produce of the Tirolean pastures to Vienna. Oberau is situated on a commanding *plateau*, and its unpretending inn '*auf* dem Keller,' offers a good resting-place. The church was burnt down in 1719, and the present one, remarkable for its size if for nothing else, was completed just a hundred years ago. It is, however, remarkable also for its altar-piece — the Blessed Virgin between S. Barbara and S. Margaret—by a local artist, and far above what might be expected in so sequestered a situation. At a distance of three or four miles, Niederau is reached, passing first a sulphur spring, esteemed by the peasants of the neighbourhood. The openest and most smiling—most *friendly*, to use the German expression —part of the valley is between Auffach and Kelchsau, where is situated Kobach, near which may be seen lateral shafts of the old mines extending to a distance of many hundred feet. From Kelchsau a foot-path leads in an hour more to Hörbrunn, where there is a brisk little establishment of glass-works, whose productions go all over Tirol. Then westwards over the Plaknerjoch to Altbach, passing Thierberg (not the same as that mentioned near Kufstein), once the chief seat of the silver-works, its only remaining attraction being the beautiful view to be obtained from its heights over the banks of the Inn, and the whole extent of country between it and Bavaria. From Altbach it is an hour more back to Brixlegg.

The memory of the former metallic wealth of the valley is preserved in numerous tales of sudden riches overtaking the people in all manner of different ways, as in the specimens already given. Here is a similar one belonging to this spot. A peasant going out with his waggon found one day in the way a heap of fine white wheat. Shocked that God's precious gift should be trodden under foot, he stopped his team and gathered up the grain, of which there was more than enough to fill all his pockets; when he arrived at his destination, he found them full of glittering pieces of money. The origin of the story doubtless may be traced to some lucky take of ore which the finder was able to sell at the market town; and the price which he brought home was spoken of as the actual article discovered. Another relic of the mining works may perhaps be found in the following instance of another class of stories, though some very like it doubtless refer to an earlier belief in hobgoblins closely allied to our own Robin Goodfellow. I think a large number date from occasions when the *Knappen* or miners, who formed a tribe apart, may have come to the aid of the country people when in difficulty.

The Unterhausberg family was once powerful in Wiltschnau. When their mighty house was building, the great foundation-stone was so ponderous that it defied all the efforts of the builders to put it in its place. At last they sat down to dinner; then there suddenly came out of the mountain side a number of Wiltschnau dwarfs, who, without any effort, lowered the great stone into its appointed place; the men offered them the best portion of their dinner, but they refused any reward. The dwarfs were not always so urbane, however, and there are many stories of their tricks: lying down in the pathways in the dark to make the people tumble over them; then hiding behind a tree, and with loud laughter mocking the disaster;[1] throwing handfuls of pebbles and ashes at the peasant girls as they passed; getting into the store-room, and mixing together the potatoes, carrots, grain, and flour, which the housewife had carefully assorted and arranged. It was particularly on women that their tricks were played off; and this to such an extent that it became the custom, even now prevailing, never to send women to the Hochalm with the herds, though they go out into other equally remote mountain districts without fear, for their *Kasa* (the hut for shelter at night, here so called, in other parts *Sennhütte*,) was sure to be beset with the dwarfs, and their milk-pails overturned. All these feats may, I think, be ascribed in their origin to the *Knappen*.

The neighbourhood of Thierberg has a story which I think also has its source in mining memories. 'On the way between Altbach and Thierbach you pass two houses bearing the name of "*beim Thaler.*" In olden time there lived here a peasant of moderate means, who owned several head of cattle; Moidl, the maid, whose

---

[1] The merry mocking laugh was a distinguishing characteristic of Robin Goodfellow. 'Mr. Launcelot Mirehouse, Rector of Pestwood, Wilts, did aver to me, *super verbum sacerdotis*, that he did once heare such a lowd laugh on the other side of a hedge, and was sure that no human lungs could afford such a laugh.'—John Aubrey, in Thoms' *Anecdotes and Traditions*, Camb. Camden Society, 1839.

duty it was to take them out to pasture on the sunny hill-side, always looked out anxiously for the first tokens of spring; for she loved better to watch the cows and goats browsing the fresh grass, or venturously climbing the heights, to sitting in the chimney-corner dozing over the spinning-wheel. One day as she was at her favourite occupation, she heard a noise behind her, and turning round saw a door open in the mountain side, and two or three little men with long beards peeping out. Within, all was dazzling with gold like the brightest sunshine. The walls were covered with plates of gold, placed one over the other like scales, and knobs of gold like pine-apples studded the vault. The little men beckoned to Moidl to come in, but she, like a modest maiden, ran home to her father; when he returned with her, however, to the spot, the door was no more to be found.' I think it may very well be imagined that Moidl came unawares upon the opening of a lateral shaft, and listened to the accounts which the Knappen may have amused themselves with giving her of the riches of their diggings; while she may very naturally have been afraid to explore these. The disappearance of the mysterious opening is but the ordinary *refrain* of marvellous tales.

The Witschnauers cannot be accused of any dreamy longings after the recurrence of such prosperous times. They are among the most diligent tillers of the land to be found anywhere; the plough is carried over places where the uneven gradients make the guiding of horses or oxen a too great expenditure of time; in such places they do not disdain to harness themselves to the plough, and even the women take their turn in relieving them. Of one husbandman of olden time it is narrated that he was even *too* eager in his thrift, and carried his furrow a little way on to his neighbour's land year by year, so that by the time he came to die he had appropriated a good strip of land not his own. His penance was, that after death he should continually tread up and down the stolen soil, dragging after him a red-hot ploughshare, in performing which his wail was often overhead—

> O weh! wie is der Pflug so heiss
> Und niemand mir zu helfen weiss! [1]

until one of his successors in the farm, being a particularly honourable man, removed the boundary-stone back to its original position. He had no sooner done so than he had the satisfaction of hearing the spectre cry—

> Erlöst, Gott sei Dank, bin ich jetzt
> Der Markstein ist auch rechtgesetzt. [2]

Another class of legends has also a home in this locality. It is told that a peasant from Oberau was going home from Thierbach, one Epiphany Eve. It was a cold night; his feet crunched the crisp snow at every step; the air was clear, and the stars shone brightly.

---

[1] O woe! the plough like fire glows,
And no one how to help me knows.

[2] Released am I now, God be praised,
And the bound-stone again rightly placed.

The peasant's head, however, was not so clear as the sky, for he came from the tavern, where he had been spending a merry evening with his boon companions. Thus it happened that instead of walking straight on, he gave one backward step for every three forward, like the *Umgehende Schuster*;[1] and thus he went staggering about till he came to the *Rastbank*, which is even yet sought as a point where to rest and overlook the view. It struck twelve as he seated himself on the bench; then suddenly behind him he heard a sound of many voices, which came on nearer and nearer, and then *the Berchtl* in her white clothing, her broken ploughshare in her hand, and all her train of little people [2] swept clattering and chattering close past him. The least was the last, and it wore a long shirt which got in the way of its little bare feet, and kept tripping it up. The peasant had sense enough left to feel compassion, so he took his garter off and bound it for a girdle round the infant, and then set it again on its way. When the Berchtl saw what he had done, she turned back and thanked him, and told him that in return for his compassion his children should never come to want. This story, I think there is little doubt, may be genuine; your Wiltschenauer is as fond of brandy as your Zillerthaler, and under its influence the peasant may very

---

[1] The haunting cobbler—a popular name for 'the wandering Jew;' in Switzerland they call him 'Der Umgehende Jud.'

[2] (The souls of all unbaptized children.) Börner, *Volkssagen*, p. 133.

likely have passed a troubled night on the Rastbank. What more likely to cross his fancy on the Epiphany Eve than the thought of a visit from the Berchtl and her children (they always appear in Tirol at that season, and in rags and tatters[1]); his own temperament being compassionate, that he should help the stumbling little one, and that the Berchtl should give him promise of reward was all that might be expected from certain premises. But what are those premises? Who was the Berchtl? If you ask a Tirolean peasant the question, he will probably tell you that the *Perchtl* (as he will call her) is Pontius Pilate's wife,[2] to whom redemption was given by reason of her intervention in favour of the Man of Sorrows, but that it is her penance to wander over the earth till the last day as a restless spirit; and that as the Epiphany was the season of favour to the Gentiles, among whose first-fruits she was, it is at that season she is most often seen, and in her most favourable mood. It must be confessed that some of his stories of her will betray a certain amount of inconsistency, for he will represent her carrying off children, wounding belated passengers, and performing many acts inconsistent with the character of a penitent soul, and more in accordance with that of the more ancient 'Lamia.'

If you address your question to Grimm, or Wolf,

---

[1] A precisely similar superstition is mentioned in Mrs. Whitcomb's recently published volume as existing in Devonshire. We shall meet Berchtl again in the neighbouring 'Gebiet der Grossen Ache' on our excursion from 'Wörgl to Vienna.'

[2] Procula is the name given her in the Apocryphal Gospels.

Simrock, Kuhn, Schwartz, or Mannhardt, or any who have made comparative mythology their study, he will tell you that the stories about her (and probably all the other marvellous tales of the people also) are to be traced back to the earliest mythological traditions of a primeval glimmering of religion spread abroad over the whole world; and to the poetical forms of expression of a primitive population describing the wonderful but constantly repeated operations of nature.[1] That the *wilder Jäger* was originally the god Wodin, the hunter of unerring aim, that his impetuous course typifies the journey of the sun-god through the heavens,[2] his mighty arm represents his powerful rays; and in even so late a tale as 'that of William Tell, he will see the last reflections of the sun-god, whether we call him Indra, or Apollo, or Ulysses.'[3] He will tell you that all 'the countless legends of princesses kept in dark prisons and invariably delivered by a young bright knight can all be traced back to mythological traditions about the spring being released from the bonds of winter; the

---

[1] 'It is now known that such tales are not the invention of individual writers, but that they are the last remnants—the *detritus*, if we may say so—of an ancient mythology; that some of the principal heroes bear the nicknames of old heathen gods; and that, in spite of the powerful dilution produced by the admixture of Christian ideas, the old leaven of heathendom can still be discovered in many stories now innocently told by German nurses, of saints, apostles, and the Virgin Mary.'—Max Müller, *Chips from a German Workshop.*

[2] Compare Cox's *Mythology of the Aryan Nations*, vol. ii. p. 864, and *passim.*

[3] Max Müller. Review of *Dasent's Works.*

sun being rescued from the darkness of night; the dawn being brought back from the far west; the waters being set free from the prison of clouds.'[1] And of the Berchtl herself, he will tell you that she is Perahta (*the bright*), daughter of Dagha (*the day*), whose name has successively been transformed into Perchtl and Bertha; brightness or whiteness has made her to be considered the goddess of winter; who particularly visited the earth for twelve winter nights, and spoilt all the flax of those idle maidens who left any unspun on the last day of the year;[2] who carries in her hand a broken plough in token that the ground is hardened against tillage; whose brightness has also made her to be reckoned the all-producing earth-mother, with golden hair like the waving corn; the Hertha of the Swabian; the Jörtha of Scandinavian;[3] the Berecynthia of the Phrygian;[4] and to other nations known as Cybele, Rhea, Isis, Diana.[5]

Such ideas were too deeply rooted in the minds of the people to be easily superseded; as my friend, the Feldkirch postilion, said, they went on and on like the echoes of their own mountains. 'The missionaries were not afraid of the old heathen gods; . . . . their kindly feeling towards the traditions, customs, and prejudices of their converts must have been beneficial; . . . .

---

[1] Max Müller. *Comparative Mythology.*
[2] A tradition still held of the Berchtl in many parts of Tirol.
[3] Nork. *Mythologie der Volkssagen.*
[4] Abbé Banier. *Mythology Explained from History.* Vol. ii. Book 3, p. 564, note *a*.
[5] Nork, Banier, &c. Cox's *Mythology of the Aryan Nations*, vol. i. pp. 317-8 and note, gives other connexions of the Legend; and at vol. ii. p. 306, and note to p. 365.

they allowed them the use of the name *Allfadir*, whom they had invoked in the prayers of their childhood, when praying to Him who is " our Father in heaven." ' And as with the greater, so with the less, the mighty powers they had personified and treated as heroes and examples lived on in their imagination, and their glorious deeds came to be ascribed to the new athletes of a brighter faith. Then, ' although originally popular tales were reproductions of more ancient legends, yet after a time a general taste was created for marvellous stories, and new ones were invented in large numbers. Even in these purely imaginative productions, analogies may be discovered with more genuine tales, because they were made after the original patterns, and in many cases were mere variations on an ancient air.' [1] More than this, there came the actual accession of marvels derived from the acts inspired by the new faith; but it cannot be denied that the two became strangely blended in the popular mind.

Brixlegg presents some appearance of thriving, through the smelting and wire-drawing works for the copper ore brought from the neighbourhood of Schwatz. It also enjoys some celebrity as the birthplace of the Tirolean historian Burgleckner, whose family had been respected here for generations; and it is very possible to put up for the night at the *Herrenhaus*. It is not much above a mile hence to Rattenberg, of which I have already spoken.

[1] M. Müller. Review of Kelley's *Indo-European Traditions*.

Rattenberg was, in 1651, the scene of a tragic event, sad as the denouement of many a fiction. The high-spirited consort of Archduke Leopold V., Claudia de' Medici, who, at his death, governed the country so well, and by her sagacity kept her dominions at peace, while the rest of Germany was immersed in the horrors of the Thirty Years' War, yet did not altogether escape the charge of occasional harshness in collecting the revenues which she knew so well how to administer. Her chancellor, Wilhelm Biener, a trusty and devoted servant and counsellor, drew on himself considerable odium for his zeal in these matters. On one occasion he got into a serious controversy with Crosini, Bishop of Brixen, concerning the payment of certain taxes from which the prelate claimed exemption, and in the course of it wrote him a letter couched in such very unguarded terms, that the bishop, unused to be so dealt with, could not forbear exclaiming, 'The man deserves to lose the fingers that could write such an intemperate effusion!' The exclamation was not thought of again till years after.

Claudia died in 1648, and then the hatred against Biener, which was also in some measure a hatred of races, for Claudia had many southerners at her court, broke forth without hindrance. He was accused [1] of

[1] Weber says the only accusation was grounded on a pasquinade against Claudia found among his papers, but that he should calumniate her seems inconsistent with his general character. Though his unsparing lampoons on his adversaries had excited them more than anything else against him.

appropriating the State money he had been so earnest in collecting, and though tried by two Italian judges, he was ultimately condemned, in 1651, to lose his head. Biener sent a statement of his case to the Archduke Ferdinand Karl; and the young prince, believing the honesty of his mother's faithful adviser, immediately ordered a reprieve. The worst enemy and prime accuser of the fallen favourite was Schmaus, President of the Council, this time a German, and he contrived by detaining the messenger to make him arrive just too late in Rattenberg, then still a strong fortress, where he lay confined, and where the sentence was to be carried out.

Biener had all along steadfastly maintained his innocence; and stepping on to the scaffold, he had again repeated the assertion, adding, 'So truly as I am innocent, I summon my accuser before the Judgment-seat above before another year is out.'[1] When the executioner stooped to lift up the head before the people, he found lying by its side three fingers of his right hand, without having had any knowledge that he had struck them off, though he might have done so by the unhappy man having raised his hand in the way of the sword in the last struggle. The people, however, saw in it the fulfilment of the words of the bishop, as well as a ghastly challenge accompanying his dying message to President Schmaus. Nor did they forget to note that the latter died of a terrible malady some months before the close

[1] Compare Gebhart, vol. ii. p. 240.

of the year. Biener's wife lost her senses when she knew the terrible circumstances of his death; the consolations of her director and of her son, who lived to his ninetieth year in the Francescan convent at Innsbruck, were alike powerless to calm her. She escaped in the night, and wandered out into the mountains no one knows whither. But the people say she lives on to be a witness of her husband's innocence, and may be met on lonely ways proclaiming it, but never harming any. Only, when anyone is to die in Büchsenhausen,[1] where her married life passed so pleasantly, the 'Bienerweible' will appear and warn them. It is a remarkable instance of the easy way in which one myth passes into another, that though this event happened but a little over two hundred years ago, the Bienerweible and the Berchtl are already confounded in the popular mind.[2]

Another name prized in Tirolese annals, which must not be forgotten in connexion with Rattenberg, is Alois Sandbichler, the Bible commentator, who was born there in 1751. He passed a brilliant career as Professor in the University of Salzburg, but died at the age of eighty in his native village.

The neighbourhood of Brixlegg is very pretty, and the views from the bridge by no means to be overlooked.

---

[1] Near Innsbruck.
[2] Staffler, *Das Deutsche Tirol*, vol. i. p. 751; and Thaler, *Geschichte Tirols v. der Urzeit*, p. 279.

CHAPTER V.

NORTH TIROL.—UNTERINNTHAL.
(LEFT INN-BANK.)

The hilles, where dwelled holy saintes,
I reverence and adore
Not for themselfe but for the sainets
Which han been dead of yore.
And now they been to heaven forewent,
Their good is with them goe;
Their sample onely to us lent,
That als we mought doe soe.—SPENSER.

WE have hitherto been occupied almost exclusively with the right bank of the Inn. We will now return to Jenbach, as a starting-point for the beauties of the left bank.

Near the station of Jenbach is a '*Restauration*,' which bears the singular title of '*zum Tolerantz*.' In the town, which is at some little distance on the Käsbach stream, the 'Post' affords very decent accommodation: The dining-room of the more primitive 'Brau' is a neat building in the Swiss style, and commands a prospect which might more than compensate for even worse fare than it affords. Jenbach had its name from being situated on the further side of the Inn from that on which the old post-road had been carried. There are extensive iron-foundries and breweries, which give the place a busy aspect, and an air of prosperity.

The excursions from Jenbach are countless. Between the stations of Brixlegg and Jenbach lie only Münster and Wiesing, with nothing remarkable, except that the church of Wiesing, having been struck by lightning in 1782, was rebuilt with stones taken from the neighbouring Pulverthurm, built by the Emperor Maximilian, in 1504, but destroyed by lightning at the same time as the church. Count Tannenberg's park (*Thiergarten*), near here, is a most curious enclosure of natural rock, aided by masonry, and stocked with deer, fish, and fowl. Then Kramsach, and in the woods near it the Hilariusbergl, once inhabited by two hermits, and still held sacred: also the strangely wild Rettengschöss and its marbles; and several remarkable Alpine peaks, particularly the Zireinalpe and its little lake, bearing a memory of Seirens in its traditions as well as in its name. Here another river Ache runs into the Inn, distinguished from that on the opposite side, as the Brandenberger Ache. At its debouche stands Voldepp, whence the Mariathal and the Mooserthal may be visited, and 'the neighbourhood is rich in marbles used in the churches of Innsbruck.'[1] The Mooserthal is remarkable for three small lakes, which can be formed and let off at pleasure; they are the property of the Barons of Lichtenthurm, who fatten

[1] Ball's Central Alps.

carp in them. The lowest of the three, the Rheinthalersee, has the prettiest surroundings. Weber says they are all fed by subterranean currents from the mountains. Ball ('Central Alps') treats them as overflowings of the Inn.

The most flourishing town of the Mariathal is Achenrain, where there are extensive brass-works. Mass is said for the out-lying operatives in the Castle-chapel of Lichtenthurm. The village of Mariathal is very snugly situated, almost hidden by its woods from the road. Its chief feature is the deserted convent of Dominicanesses founded in the thirteenth century by Ulrich and Konrad v. Freundsberg; their descendant, Georg v. Freundsberg, celebrated in the Thirty Years' War, whom we learn more about when we come to Schwatz, also endowed the nuns liberally, bidding them pray for him; his effigy may still be seen in the church of Mariathal; and the convent, even in its present condition, is a favourite pilgrimage. Hence a rocky defile of wild and varied beauty, and many miles in length, leads into the Brandenbergerthal, which reaches to the Bavarian frontier. Its highest point is the Steinberg, to be recognized in the distance by its pyramidal form, which is situated within what the Germans graphically term a *cauldron* (Gebirgskessel) of mountains, and is shut off from all communication with the outer world by the snow during the winter months. The Brandenbergers have been famous for their patriotism and defence of their independence during all the various conflicts with Bavaria, and they love to call their native soil the *Heimaththal* and the *Freiheitthal*. The only tale of the supernatural I have met with as connected with this locality is the following; it has a certain wild grasp, but its moral is not easy to trace; it is analogous, however, to many traditions of other places.

'One of the Jochs surrounding the Brandenbergerthal was celebrated for its rich grasses; on its "alm"[1] the cattle often found pasturage even late in the winter. The *Senner*[2] here watching his flocks was visited one Christmas Eve by an old man in thick winter clothing, with a mighty pine-staff in his hand; he begged the *Senner* on the coming night to heat his hut as hot as ever he could, assuring him he would have no cause to regret his compliance. The *Senner* thought it was a strange adventure, but congratulated himself that it might be the means of propitiating the goblins, of whose pranks in the winter nights he was not without his fears. So he heaped log upon log all day, till the hut was so hot he could hardly bear it. Then he crept under a bench in the corner where a little chink gave a breath from the outer air, and waited to see what would come to pass. Towards midnight he heard steps approaching nearer and nearer, and then there was a sound of heavy boots stamping off the snow. Immediately after, seven men stepped

---

[1] Pasture-ground lying at the base of a mountain.
[2] Alpine herdsman.

into the room in silence. Their boots and clothes were all frozen as hard as if they had been carved out of ice, and their very presence served to cool down the air of the hut to such an extent that the *Senner* was now obliged to rub his hands. When they had stood a considerable space round the fire without uttering a word, they all seven left the hut as silently and solemnly as they had entered it. The *Senner* now crawled out of his hiding-place, and a loud cry of joy burst spontaneously from his lips, for his hat, which he had left on the table, was full of bright shining golden *zwanzigers*. These seven,' the legend goes on to say, 'were never seen but this once. They were the seven *Goldherds* of the *Reiche Spitze* (on the Salzburg frontier); for up there there are exhaustless treasures, but whatever a mortal takes of them during life, he must suffer *the Cold Torment* and keep watch over it after death ; and of such there have been seven in the course of the world's ages.'

With regard to 'the Cold Torment,'[1] they have the following legend in the neighbourhood of Innsbruck :—There was once a peasant who had been very unlucky, and got so deep in debt that he saw no way of extricating himself. Unable to bear the sight of his starving family, he wandered out into the forest,

---

[1] Respecting the curious idea of the *kalte Pein*, consult Alpenburg, *Mythen Tirols.* ; Vernalken, *Alpensagen ;* Beckstein, *Thuringer Sagenbuch.* See also Dr. Dasent's remarks about Hel in *Popular Tales from the Norse* ; and Dante (notably *Inferno*, cantos vi. xxii. xxiv.) introduces cold among the pains of even the Christian idea of future punishment.

until at last he met a strange-looking man in the old Frankish costume, who came up to him and said, ' You are poor indeed, and know no means of help.' 'Most true,' replied the peasant; 'of money and good counsel I can use more than you can have to bestow.' 'I will help you,' said the strange-looking man; 'I will give you as much money as you can use while you live, and all you have to do for it will be to bear *the Cold Torment* for me after you die; nothing but that, only just to feel rather too cold, and all that time hence—what does it matter?' The peasant retraced his steps, and as he drew near home his children came out to meet him with their pinafores full of gold, and all about the house there were heaps of gold, more than he could use; and he lived a merry life till the time came for him to die. Then he remembered what was before him; so he called his wife to him, and got her to make him a whole suit of the thickest rough woollen cloth, and stockings, hood, and gloves of the same. In the night, before they had buried him, his boys saw him, just as the De profundis bell rang, get up from the bed in all this warm clothing, and shut the gate behind him, and go out into the forest to deliver the spirit which had enriched him.[1]

---

[1] Here we have quite the Etruscan idea of providing against after-death needs with appliances connected with the mortal state. Dennis (*Cities and Cemeteries of Etruria*. vol. i. p. 34) mentions more material traces of Etruscan beliefs at Matrei, on the north side of the Brenner. Somewhat further south more important remains still have of late years been unearthed, as we shall have occasion to note by-and-by.

The story in the text, in its depiction of self-devotion, has much

To the north-east of this valley, and still on the left bank of the Inn, is the favourite pilgrimage of Maria-Stein. I have not learnt its origin, but there is a tradition that, in 1587, Baron Schurff, to whom the neighbouring Castle of Stein then belonged, being desirous to take the precious likeness of the Blessed Virgin honoured there to his Bavarian dwelling, thrice attempted the removal, and on each occasion it was found by the next morning restored to its original sanctuary, which is in a chapel at the top of a high tower. The castle was a dependency of the Freundsbergers of Schwatz, till the family died out. It was subsequently bestowed by the Archduke Sigismund on one of his supporters, to whom he gave also the title of Baron Schurff. Afterwards it came into possession of Count Paris von Klotz, who gave it to form a presbytery and school for which it is still used. Among its

analogy with a Chinese legend told to me by Dr. Samuel Birch, of the British Museum, concerning a man who sacrifices his own life in order to put himself on fighting terms with a cruel spirit which torments that of his dead companion. In its details it is like the story I have pointed out in *Folklore of Rome* (the 'Tale of the Pilgrim Husband,' pp. 361-3 and xvii), as the most devious from Christian teaching of any of the legends I have met with in Rome; and it is particularly noteworthy in connexion with Mr. Isaac Taylor's summary of the Etruscan creed (*Etruscan Researches*, p. 270). 'The Turanian creed was Animistic. The gods needed no gifts, but the wants of the ancestral spirits had to be supplied : the spirits of the departed were served in the ghost-world by the spirits of the utensils and ornaments which they had used in life.') And in effect we find in every collection of the contents preserved at the opening of Etruscan tombs, not only gems and jewellery and household utensils, but remains also of every kind of food.

treasures was a Slave codex of Homilies of the early fathers; Count Klotz had a reprint made from it at Vienna. A little lake (Maria Steinersee) at no great distance affords excellent fish called *Nasen*, whence the neighbouring dale is called Nasenthal; and from several points there are most enjoyable views of the *höhe Salve* and the little towns of Wörgl, Kirchbühel, and Häring across the river.

Jenbach affords also numerous mountain walks through the Achenthal: a favourite one is over the Mauriz Alp, to Maurach, which has many points of interest to the geologist. For those who are not fond of pedestrianism, there is a splendid drive along the road—one of the old highways to Bavaria and the north of Europe. An accident is of very rare occurrence; but some parts of it are rather frightful. For those whose nerves are proof against the fears suggested here and there, there is immense enjoyment to be found, as it winds its way along the romantic woody Käsbachthal, round—indeed through— the wild and overhanging rocks, or, supported on piles, runs close along the edge of the intensely blue Achen lake, under the over-arching Spiel-joch, steep as a wall. The first place to halt at is Skolastica, where there is a pretty, much-frequented swimming-school; and whence even ladies have ascended the Unnutzjoch over the Kögl. It is often crowded in the season, as also are all the little towns round the lake—Achenthal, Pertisau,

Buchau. Several excellent varieties of fish, which are the property of the Monastery of Viecht, and the pleasure-fares across the waters, afford means of subsistence to a little population of boatmen, who have made their nests on the rocks wherever there is a foot of level ground. Pertisau, however, is on a green smiling spot, and is a relief to the majestic wildness of the rest of the surrounding scenery. A very extraordinary effect may be observed at a short distance out from Buchau. The mountain outline on the right hand appears to be that of a regular fortress, with all professional accessories, bidding defiance to the neighbourhood: it is only as the boat approaches quite near, that you see it is only one of those *tours de force* with which nature often surprises us; as, for example, in the portrait of Louis XVI. in the outline of the Traunstein, seen from Baura.

From the village of Achenthal the road runs, through the Bavarian frontier, to the well-known baths and Bavarian royal *Lustschloss*—until 1803 a Benedictine monastery—of Tegernsee, through *Pass-Achen*, celebrated in the patriotic struggles of 1809.

The Achensee is the largest and one of the most beautiful lakes of Tirol. It is fed partly by mountain streams, and partly by subterranean springs. The people tell a warning tale of its first rising. They say that in olden times there was a stately and prosperous town on what is now the bed of the lake; but the inhabitants in their prosperity forgot God so far, that the young lads played at skittles along the aisles of the church, even while the sacred office was being sung, and the Word of God preached. A day came; it was a great feast, but they drove their profane sport as usual, and no one said them nay;[1] and so a great flood rose up through the floor; rose above their heads; above the church roof; above the church steeple; and they say that even now, on a bright calm day, you may see the gilt ball of the steeple shining under the waters, and in the still moonshine you may hear the bell ring out the midnight hour. There are many other tales of such swift and righteous judgments lingering in Tirol.

The lower eastern ridge of the Harlesanger or Hornanger Alpe, is, on account of its stern and barren character, called the Wildenfeld. This is how it received its name. Ages ago, it was a very paradise of beauty and fruitfulness. All the choicest Alpine grasses grew

---

[1] There is something like this in Dean Milman's *Annals of St. Paul's Cathedral*:—' "Others," adds Bishop Braybroke, "by the instigation of the devil, do not scruple to play at ball, and other unseemly games, within the church (he is speaking of St. Paul's), breaking the costly painted windows, to the amazement of the spectators."' Speaking of the post-Reformation period, the Dean adds: 'If, when the cathedral was more or less occupied by sacred subjects, the invasion of the sanctuary by worldly sinners resisted all attempts at suppression; now, that the daily service had shrunk into mere forms of prayer, at best into a mere "Cathedral Service," . . . . . it cannot be wondered at that the reverence, which all the splendour of the old ritual could not maintain, died away altogether as Puritanism rose in the ascendant.' Mr. Longman, however (*The Three Cathedrals*, p. 54-6), quotes the very stringent regulations which were issued for the repression of such practices: perhaps the legend constructor would say, these afford the reason why, though St. Paul's was profaned like the church of Achensee, it did not 'likewise perish.'

there in abundance; but with these riches and plenty the pride of the Senners and milkers waxed great too; and as a token of their reckless wastefulness, it is recorded that they used rich cheeses for paving-stones and skittles. One ancient Senner, like another Lot, raised his feeble but indignant voice against them, but they heeded him not. One day, as he mused over the sins of his people, a bright bird, with a plumage such as he had never seen before, fluttered round him, warbling, 'Righteous man, get thee hence! righteous man, get thee hence!' The old man saw the finger of God, and immediately followed the guiding flight of the bird to a place of safety, while a great peak from the Harlesanger fell over the too prosperous Joch, buried its impious inhabitants, and spread desolation all around. There is now a pilgrimage chapel.

Another excursion, which must not be omitted, from Jenbach, is that to Eben, which lies a little off the high road, at some elevation, but in the midst of a delightful table-land (hence its name) of most fruitful character. As the burial-place of St. Nothburga, it is still a spot of great resort. Unhappily, not all those buried here were so holy as the peasant saint. A tradition is preserved of one wicked above others, though he seemed all fair to the outward eye, and the Church consequently admitted him to lie in holy ground. But he felt the Eye of One above upon him, and he could not rest; and in his struggles to withdraw himself from that all-searching gaze, he bored and bored on through the consecrated earth, till he had worked his way out into the common soil beyond. A horse-shoe, deeply graven in the 'Friedhof' boundary, and which no one has ever been able to wall up, marks the spot by which he passed: and the people call it the 'Escape of the Vampire.'[1]

The unpretending village of Stans, situated in the midst of a very forest of fruit-trees, at no great distance from Jenbach, is the birth-place of Joseph Arnold, one of the religious artists, of whom Tirol has produced so many. Without winning, of some it may be said without meriting perhaps, much fame for themselves in the world, without attaining the honour of founding a school, they have laboured painstakingly and successfully to adorn their village temples, and keep alive the faith and devotion of their countrymen. Almost wherever you go in Tirol you find praiseworthy copies of paintings, whose titles are connected with the celebrated shrines of Italy, modestly reproduced by them, or some fervent attempt at an original rendering of a sacred subject, by men who never aspired that their names should reach beyond the echoes of their own beloved mountains. The prior of Viecht, Eberhard Zobel, discovered the merits of Joseph Arnold and drew him from obscurity, or rather from one degree of obscurity

---

[1] Nork (*Mythologie der Volksagen*, vol. ix. p. 83) gives other significations to horse-shoes found in the walls of old churches, but does not mention this instance. Concerning the origin of the superstition about vampires, see Cox's *Mythology of the Aryan Nations*, vol. i. p. 363; also p. 63 and p. 429.

to another less profound, had him instructed according to the best means within his attainment, and gave him occupation in the monastery. His homely aspirations made him content with the sphere to which he was native, and he never went far from it. The altarpiece in the church of Stans, representing St. Lawrence and St. Ulric, is his work and his gift.

From Stans there is a path through the grand scenery of the Stallenthal, leading to the shrine of St. Georgenberg. For a time the pretty villages of the Innthal are lost to sight, and you pass a country known only to the wild game, the hunter, and the pilgrim; the bare rocky precipices relieved only here and there with woods, while the Stallen torrents run noisily below. Who could pass through such a neighbourhood and not think of the crowds of pilgrims who, through ages past, have approached the sacred spot in a spirit of faith and submission, bearing their sins and their sorrows, the burden of their afflictions, moral and physical, and have gone down to their homes comforted?

A wonderful shrine it is: a rock which might seem marked out 'from the beginning' to be a shrine; shut out by Nature from earthly communication; piercing the very sky. You stand beneath it and long for an eagle's wings to bear you aloft: there seems no other means of access. Then a weary winding path is shown you, up which, with many sloping returns upon your former level, and crossing the roaring stream at a giddy height, you at last reach an *Absatzbrücke*—a bridge or viaduct—over the chasm, uniting the height you have been climbing, with the cliff of S. George. It is a long bridge, and only made of wood; and you fancy it trembles beneath your anxious tread, as you span the seemingly unfathomable abyss. A modest cross, which you cannot fail to observe at its head, records the marvellous preservation of a girl of twenty-one, named Monica Ragel, a farm-servant, who one fine morning in April 1831, in her zeal to gather the fairest flowers for the wreath she was weaving for the Madonna's altar, attempted to climb the treacherous steep, and losing her footing slipped down the cliff, a distance of one hundred and forty feet. The neighbours crowded to the spot, with all the haste the dangerous footing would admit, and though they had no hope of finding her alive. She was so far uninjured, however, that she was able to resume work within the week.

The buildings found perched at this height cannot fail to convey a striking impression; and this still more do the earnest penitents, who may nearly always be found kneeling within. First, you come upon the little chapel of the '*Schmerzhaften Mutter*,' with a little garden of graves of those who have longed to lie in death as they dwelt in life—near the shrine; among them is that of the Benedictine Magnus Dagn, whose knowledge of music is referred to in the following simple epitaph, '*Magnus nomine, major arte, maximus virtute.*' Opposite it is the principal church, containing in one of its chapels one of those most strange of relics, which here and there have come down to us with their legends from 'the ages of faith.' In the year 1310, when Rupert I. was the fourteenth abbot of St. Georgenberg, a priest of the order[1] was saying Mass in this very chapel. Just at the moment of the consecration of the chalice a doubt started in his mind, whether it were possible that at his unworthy bidding so great a mystery should be accomplished as the fulfillment of the high announcement, 'This is My Blood.' In this condition of mind he concluded the words of consecration; and behold, immediately, in place of the white wine mingled with water in the chalice, he saw it fill with red blood, overflowing upon the corporal; some portion of this was preserved in a vial, set into a reliquary on the altar. Round the church are the remains of the original monastery, in which the monks of Veicht generally leave some of their number to minister both to the spiritual and corporal needs of pilgrims.

It seems difficult to fix a date for the origin of this pilgrimage, one of the most ancient of Tirol. There is a record that in 992 a chapel was consecrated here to our Lady of Sorrows, by Albuin, Bishop of Brizen; but it was long before this[2] that Rathold, a young nobleman of Aiblingen in Bavaria, 'having learnt the hollowness of the joys his position promised him, made up his

---

[1] Gebhart.   [2] 'Probably early in the ninth century.'—*Scherer.*

mind to forsake all, and live in the wilderness to God alone.' He wandered on, shunning the smooth and verdant plains of his native lands, and the smiling fruitful amenities of the Innthal, till at last he found himself surrounded by wild solitudes in the valley of the Stallen; plunging into its depths, his eye alighted on the almost inaccessible Lampsenjock. Then choosing for his dwelling a peak, on which a few limes had found a ledge and sown themselves, he cut a little cave for his shelter in the rock beneath them, and there he lived and prayed. But after a time a desire came over him to visit the shrines of the mightiest saints; so he took up his pilgrim staff once more, and sped over the mountains and over the plains, till he had knelt at the *limine Apostolorum*, and pressed his lips upon the soil, fragrant with the martyr's blood. Nor was his zeal yet satisfied. There was another Apostle the fame of whose shrine was great; and 'a year and a day' brought our pilgrim to S. Iago de Compostella. Then, having thus graduated in the school of the saints, he came back to his solitude under the lime-trees on the rock, to practise the lessons of Divine contemplation he had thus imbibed in the perfume of the holy places.

He did not come back alone. From the great storehouse of Rome he had brought a treasure of sacred art—a picture of the Madonna, for which his own hands wrought a little sanctuary. From far and near pious people came to venerate the sacred image; and ' *Unsere liebe Frau zur Linde*,' was the watch-word, at sound of which the sick and the oppressed revived with hope.

One day, it chanced that a young noble, whom ardent love of the chase had led into this secluded valley, turned aside from following the wild chamois, to inquire what strange power fascinated the peasants into attempting yon steep ascent. Curious himself to see the wonder-working shrine, he scaled the peak, and found to his astonishment, in the modest guardian of the picture, the elder brother who long ago had 'chosen the better part.' In token of his joy at the meeting, he made a vow to build on the spot a chapel, as well as a place of shelter for the weary pilgrim. His undertaking was no sooner known than all the people of the neighbouring valleys, nobles and peasants, applied to have their part in the work. Thus supported, it was begun in right earnest; but the workmen had no sooner got it fairly in hand than all the blessing, which for so long had been poured out on the spot, seemed suddenly to be quenched. Nothing would succeed, and every attempt was baffled; and one thing, which was more particularly remarked, was that the men were continually having accidents, and wounding themselves with their tools. More strange still, every day two white doves flew down from above, and carefully picking out every chip and shaving on which blood had fallen, gathered them in their beaks and flew away. Finding that no progress could be made with the work, and that this manœuvre of the doves continued day by day, the pious Reinhold resolved to follow them; and when he at last succeeded in finding their hiding-place, there lay before him, neatly fashioned out of the chips which the doves had carried away, a tiny chapel of perfectly symmetrical form.[1] The hermit saw in the affair the guiding hand of God, demanding of him the sacrifice of seven years' attachment to his cell; and cheerfully yielding obedience to the token, requested his brother that the chapel should be erected on the spot thus pointed out. Theobald willingly complied, and dedicated it to the patron of chivalry, St. George. The fame of Reinhold's piety, and of his wonderful chapel, was bruited far and near; and now, not all who came to visit him went back to their homes. Many youths of high degree, fired by the example of the hermit sprung out of their order, applied to join him in his life of austerity; and soon a whole colony had established itself, Camaldolese-fashion, in little huts round his. There seems to have been no lack of zealous followers to sustain the odour of sanctity of St. Georgenberg; early in the twelfth century, the Bishop of Brixen put them under the rule of S. Benedict, to whose monks Tirol, and especially Unterinnthal, already owed so great a debt of gratitude, for keeping alive the faith. His followers endowed it with much of the surrounding land, which the brothers, by hard manual labour, brought into cultivation. They were overtaken by many heavy trials in the course of cen-

---

[1] Burglechner. Pilger durch Tirol. Panzer. Mülhenhof.

turies: at one time it was a fire, driven by the fierce winds, which ravaged their homestead; at another time, avalanches annihilated the traces of their industry. At last, the spirit of prudence prevailing on their earlier energetic hardiness, it was resolved to remove the monastery to Viecht, where the brothers already had a nucleus in a little hospital for the sick among them, and where also was the depôt for their cattle-dealing—a *Viehzuchthof*,[1] whence by corruption it derived its name.

The execution of this idea was commenced in 1705. The abbot, Celestin Böhmen, a native of Vienna, had formerly held a grade of officer in the Austrian artillery. Nothing could exceed the zeal with which he took the matter in hand; and plans were laid out for raising the building on the most extensive and costly scale. So grand an edifice required large funds; and these were not slow to flow in, for St. Georgenberg was beloved by all the country round. When he saw the vast sums in his hand, however, the old spirit of the world, and its covetousness, crept over him again, and a morning came when, to the astonishment of the brotherhood, the abbot was nowhere to be found—nor the gold! The progress of the work was effectually arrested for the moment; but zeal overcame even the obstacle presented by this loss, and by 1750 Abbot Lambert had brought to completion the present edifice, in late Renaissance style, which, though imposing and substantial, forms but one wing of Celestin Böhmen's plan.

[1] *Lit.* a 'cattle-breeding-farm.'

If the spirit of the world came over Abbot Celestin in the cloister, the spirit of the cloister came back upon him in the world; and it was not many years before he came back, full of shame and contrition, making open confession of his fault, and placing himself entirely in the hands of his former subjects. Though at this time the monks were yet in the midst of their anxieties for the means for carrying on the work, they suffered themselves to be ruled by a spirit of Christian charity, and refused to give him up to the rigour of the law; and he ended his days with edifying piety at Anras, in the Pusterthal.

A great festival was kept at Viecht, in 1845, in memory of the consecration, which was attended by sixty thousand persons, from Bavaria as well as Tirol.

The library contains an interesting collection of MSS. and early printed books in many languages, and is particularly rich in works illustrative of Tirolean history. In the church are some of Nissl the elder's wood-carvings, which are always worth attention. The confessionals are adorned with figures of celebrated penitents, by his hand; and other noteworthy works will be found in a series of nine *tableaux*, showing forth the Passion; also the crucifix over the high altar, and four life-sized carvings. In all these he was assisted by his pupils, Franz Thaler, of Jenbach, who afterwards came to have the charge of the Vienna cabinet of antiquities, and Antony Hüber, the most successful of his school. Perhaps the finest specimen of all is a dead Christ, under the altar, remarkable for the anatomical knowledge displayed. Like many another mountain sanctuary isolated and exposed to the wind, this monastery has more than once been ravaged by fire; in 1868 it was in great part burnt down, and the church-building zeal of Tirol is still being exercised with great energy and open-handedness in building it up again. A festival was held there in October 1870, when five bells from the foundry of Grassmayr of Wilten were set up to command the echoes of the neighbourhood; great pains are now being taken to make the building fireproof.

Close opposite Viecht lies Schwatz;[1] a number of straggling houses, called 'die lange Gasse,' on the Viecht side belong to it also; between them there is a bridge, which we will not cross now, but continue a little further along the left bank; this, though less rich in smiling pastures than the right, has many points of interest. The next village to Viecht is Vomp, situated at the entrance of the Vomperthal, the sternest and most barren in scenery or settlements of any valley of Tirol, and characterized by a hardy pedestrian as 'frightfully solitary, and difficult of

[1] It follows that (when mountain scenery is not the special object with the tourist) it is better to visit Viecht when staying at Schwatz (Chapters vi. and vii.) than from Jenbach, at least it is a much less toilsome ascent on this side from Viecht to S. Georgenberg. the most interesting point of the pilgrimage. At S. Georgenberg there is a good mountain inn.

access: even the boldest Jägers,' he adds, 'seldom pursue their game into it.' The village church of Vomp once possessed a priceless work of Albert Durer, an '*Ancona*,' showing forth in its various compartments the history of the Passion; but it was destroyed in 1809, when the French, under Deroi, set fire to the church in revenge for the havoc the Tirolean sharp-shooters had committed among their ranks. Joseph Arnold (in 1814) did his best to repair the loss, by painting another altar-piece, in which we see a less painful than the usual treatment of the martyrdom of St. Sebastian: the artist has chosen the moment at which the young warrior is being bound to the tree where he is to suffer so bravely. Above the village stands the once splendid castle of Sigmundslust, one of the hunting-seats of Sigismund the Monied (*der Münzreiche*),[1] now the villa of a private family of Innsbruck, Riccabona by name. Vomp is also the birthplace of Joseph Hell, the wood-carver.

Crossing the Vomperbach, and the fertile plain it waters, you reach Terfens, which earned some renown

---

[1] In his reign, 1440-90, it was that the silver-mines of Tirol were discovered; and the abundant influx, to the extent of 500 cwt. annually, of the precious metal into his treasury, led him to treat its stores as exhaustless; though the richest monarch of his time, his easy open-handed disposition continually led him into debt, and made his subjects finally induce him in his old age to resign in favour of his cousin, the Emperor Maximilian I. It is a token of the simplicity of the times, that one of the gravest reproaches against him was that he indulged in the luxury of *silk stockings!* He married Eleanor, daughter of James II. of Scotland.

in the wars of 'the year nine.' Outside the village is a little pilgrimage chapel, called Maria-Larch, honoured in memory of a mysterious image of the blessed Virgin, found under a larch fir on the spot, similar to the legend of that at Waldrast.[1]

Passing the ruin of Volandseck, the still inhabited castle of Thierberg (the third of the name we have passed since we entered Tirol) and the village of S. Michael, you come to S. Martin, the parish church of which owes its endowment to a hermit of modern times. There was in the village a convent, deserted, because partly destroyed by fire. In 1638, George Thaler, of Kitzbuhel, a man of some means and position, came to live here a life of sanctity: he devoted six hours a-day to prayer, six to sleep, and the rest to manual labour. He maintained a chaplain, and an old servant who waited on him for fifty years. At his death, he left all he possessed to supply the spiritual needs of the hamlet. After leaving S. Martin's, the scenery grows more pleasing: you enter the Gnadenwald, so called, because its first inhabitants were servants of the earlier princes of Tirol, who pensioned them off with holdings of the surrounding territory. It occupies the lowland bordering the river, which here widens a little, and affords in its recesses a number of the most romantic strolls. Embowered on its border, near the river, stands the village of Baumkirchen, with

---

[1] See *infra* in the Stubayerthal.

its outlying offshoot of Fritzens now surpassing it in importance, as it has been chosen for the railway-station. The advance of the iron road has not stamped out the native love for putting prominently forward the external symbols of religion. I one day saw a countryman alight here from the railway, who had been but to Innsbruck to purchase a large and handsome metal cross, to be set up in some prominent point of the village and it was considered a sufficiently important occasion for several neighbours to go out to meet him on his return with it. Again, on the newer houses, probably called into existence by the increased traffic, the old custom of adorning the exterior with frescoes of sacred subjects is well kept up. This is indeed the case on many other parts of the line; but at Fritzens, I was particularly struck with one of unusual merit, both in its execution and its adaptation to the domestic scene it was to sanctify. I would call the attention of any traveller, who has time to stop at Fritzens to see it: the treatment suggests that I should give it the title of 'the Holy Family *at home*,' so completely has the artist realized the lowly life of the earthly parents of the Saviour, and may it not be a comfort to the peasant artizan to see before his eyes the very picture of his daily toil sanctified in its exercise by the hands of Him he so specially reveres?

An analogous incident, which I observed on another occasion, comes back to my memory: it happened, I think, one day at Jenbach. The train stopped to set down a Sister of Charity, who had come to nurse some sick person in the village. The ticket-collector, who was also pointsman, was so much occupied with his deferential bowing to her as he took her ticket, that he had to rush to his points 'like mad,' or his reverent feelings might have had serious consequences for the train! So religious indeed is your whole *entourage* while in Tirol, that I have remarked when travelling through just this part in the winter season, that the very masses of frozen water, arrested by the frost as they rush down the railway cuttings and embankments, assumed in the half-light such forms as Doré might give to prostrate spectres doing penance. The foot-path on to Hall leads through a continuance of the same diversified and well-wooded scenery we have been traversing hitherto; but if time presses, it is well to take the railway for this stage, and make Hall or Innsbruck a starting-point for visiting the intervening places.

Hall is the busiest and most business-like place we have come to yet, and the first whose smoky atmosphere reminds us of home. There is not much to choose between its two inns the 'Schwarzer Bär' and the 'Schwarzer Adler.' The industry and the smoke of Hall arises from the salt-works, from which Weber also derives its name (from $ἁλός$, salt; though *why* it should have been derived from the Greek he does not explain). The first effect which strikes you on arriving, after the smokiness, is the sky-line of its bizarrely-picturesque steeples, among the most bizarre of which is the

Münzthurm (the mint-tower), first raised to turn into money the over-flowing silver stores of Sigismund *the Monied*; and last used to coin the *Sandwirthszwänziger*, the pieces of honest old Hofer's brief but triumphant dictatorship. The town has in course of time suffered severely from various calamities : fire, war, pestilence, inundation, and, on one occasion, in 1670, even from earthquake; the church tower was so severely shaken, that the watchman on its parapet was thrown to the ground; the people fled from their houses into the fields, where the Jesuit fathers stood addressing them, in preparation for their last end, which seemed imminent. Loss of life was, however, small; nevertheless, the Offices of the Church were for a long time held in the open air. Notwithstanding all these reverses, the trade in salt, and the advantageous municipal rights granted them in earlier times, have always enabled the people to recover and maintain their prosperity. In the various wars, they have borne their part with signal honour. One of their greatest feats, perhaps, occurred on May 29, 1809. Speckbacher had led his men to a gallant attack on the Bavarians at Volders, blowing up the bridge behind him, and then marched to the relief of Hall; the Bavarians were in possession of the town and bridge, and as they had several pieces of artillery, it was not easy for the patriots to carry it; nevertheless, as their ammunition was failing, and Speckbacher having refused to agree to a truce, because he saw the advantage accruing to him through this deficiency, they destroyed the Hall bridge, as they thought, and retreated homewards under cover of the night. Speckbacher discovered their flight early in the morning, and lost no time in addressing his men on the importance of at once taking possession of their native town : the men were as usual at one with him, and not one shrank from the perilous enterprise of regaining the left bank by such means as the tottering remains of the bridge afforded!

Joseph Speckbacher, who shares with Andreas Hofer the glories of 'the year nine,' was a native of Rinn, a village on the opposite bank ; but he is honoured with a grave in the Pfarrkirche, at Hall, bearing the following inscription, with the date of his death, 1820 :

> Im Kampfe wild, doch menschlich ;
> In Frieden still und den Gesetzen treu ;
> War er als Krieger, Unterthan und Mensch,
> Der Ehre wie der Liebe werth. [1]

Another object of interest, in the same churchyard, is a wooden crucifix, carved by Joseph Stocker in 1691 ; as well as the monuments of the Fiegers, and other high families of the middle ages. In the church itself is a 'Salvator Mundi' of Albert Durer, on panel ; the altar-piece of the high altar is by Erasmus Quillinus, a pupil of Rubens. One of the chapels, the Waldaufische Kapelle, was built in 1493-5, by one Florian von

---

[1] In battle impetuous, yet merciful ; in time of peace tranquil, and faithful to his country's laws : whether as a warrior, a subject, or an individual, worthy of honour as of love.

Waldauf, to whom an eventful history attaches. He was a peasant boy, whom his father's severity drove away from home : for a long time he maintained himself by tending herds; after that he went for a soldier in the Imperial army, where his talents brought him under the special notice of the Emperor Frederick, and his son Maximilian I., who took him into their councils and companionship. Maximilian made him knight of Waldenstein, and gifted him with lands and revenues. His love of adventure took him into many countries. On one journey, being in a storm at sea, the memory of his early wilfulness overcame him, and he vowed that if he came safe to land, he would build a chapel in his Tirolean home. He subsequently fixed on the Pfarrkirche of Hall, as that in which to fulfil his vow, being the parish church of the castle of Rettenberg which Maximilian had bestowed on him, and enriched it with a wondrous store of relics, which he had collected in his journeyings. Above 40,000 pilgrims flocked from every part of Tirol, to assist at the consecration; and a goodly sight it must have been, when singing and bearing the relics aloft, they streamed down the mountain side and across the river, the last of the procession not having yet left the gates of Castle Rettenberg, while the foremost had already reached the chapel.

There are other churches in Hall; where that of S. Saviour now stands was once a group of crazy cottages; but one day, in the year 1406, in one of them a poor man lay dying, and the priest bore him the holy

Viaticum, which knows no distinction between the palace and the hovel: the furniture was as rickety as the tenements themselves; the only table, on which the priest had deposited the sacred vessels, propped against the wall for support, gave way by some accident, and the *Santissimo* was thrown upon the floor. Johann von Kripp, a wealthy burgher, hearing of what had befallen, bought the cottages, and in reparation for the desecration, built a church on the spot, with the dedication, *zum Erlöser*.

The town is well provided with educational and charitable institutions; the latter comprising a madhouse worth seeing, under Professor Kaplan, and a deaf and dumb school. The Franciscan monastery is, I think, the only unsuppressed religious house. In the *Rathhaus* is preserved a quaint old picture, representing the Emperor Sigismund, in hunting costume, coming to ask the assistance of the men of Hall against a conspiracy he had discovered in Innsbruck, assistance which loyal Hall was not slow to supply. Its situation made it a place of some importance to the defences of the country; and the regulations for calling the inhabitants under arms were very complete, so that this service was promptly rendered.

An amusing story is told in evidence of the ready gallantry of the men of Hall. There was a time when Hall was at feud with the neighbouring village of Taur: the watchman, stationed on the tower by night-time, rang the alarm, and announced that the enemy was advancing with lanterns in their hands; at the call to arms, every man jumped from his bed, and seized his weapon, eager to display his prowess against the foe. Prudent *Salzmair*[1] Zott, anxious to spare the shedding of neighbours' blood, hastily donned a shirt of mail over his more penetrable night-gear, and proposed to ride out alone with a flag of truce, to know what meant the unseasonable attack. The warlike burghers with difficulty yielded to his representations, and not having the consolations of the fragrant weed wherewith to wile away their time, set to sharpening their swords and axes, and outvieing each other with many a fierce boast during his absence.

Meantime, *Salzmair* Zott proceeded on his way without meeting the ghost of a foe, or one ray from their lanterns, till he came to Taur itself, where everything lay buried in peaceful silence. Only as he came back he discovered what had given rise to the alarm: it was midsummer-tide, and a swarm of *little worms of St. John*[2] was soaring and fluttering over the fields like a troop provided with lanterns. So with a hearty laugh he despatched the townsmen, ready for the fight, back to their beds. And even now this humorous imitation of the *Bauernkrieg*[2] is a by-word for Quixotic enterprises.

Of all the numerous excursions round Hall, the

---

[1] Steward of the salt-mines.   [2] *Johanniswürmchen*, fire-flies.
[3] Peasants' war.

strangest and most interesting is that to Salzberg, the source of the salt, the crystalizing of which and despatching it all over Tirol, to Engadein and to Austria, forms the staple industry of Hall. It is a journey of about three hours, though not much over eight miles, but rugged and steep, and in some parts rather frightful, particularly in the returning descent, for the Salzberg lies 6,300 feet above the sea : but there is a road for an *einspanner* all the way; entrance is readily obtained, and the gratuities for guide, lighting up, and boat over the subterranean salt lake, exceedingly moderate. There are records extant which shew that there were salt-works in operation in the neighbourhood of Hall early in the eighth century, but these would appear to have been fed by a salt spring which flowed at the foot of the mountain. In the year 1275, however, Niklas von Rohrbach, who seems to be always styled *der fromme Ritter* (the pious knight), frequently when on his hunting expeditions in the Hallthal, observed how the cattle and wild game loved to lick certain cliffs of the valley; this led him to test the flavour, and finding it rich in salt, he followed up the track till he came to the Salzberg itself, where he prudently conjectured there was an endless supply to be obtained.[1] Ever since this time the salt has been worked pretty much in the same way, namely, by hewing, later by blasting, vast chambers in the rock, which are then filled with water and closed up : at the end of some ten or twelve

[1] Burglechner.

months, when the water is supposed to be thoroughly impregnated, it is run off through a series of conduits to Hall, where it is evaporated, a hundred pounds of brine yielding about a third the weight of salt. A considerable number of these chambers, an acre or two in extent, have been excavated in the course of time, and you are told that it would take more than a week to walk through all the passages connecting them. 'Cars filled with rubbish pass you as you thread them,' says an observant writer, 'with frightful rapidity; you step aside into a niche, and the young miners seated in the front look like gnomes directing infernal chariots.' The crystallizations in some of these chambers lighted up by the torches of a party of visitors have a magical effect, and recall the gilded fret-work of some Moorish palaces. There is a tradition that Hofer and Speckbacher, who never, before their illustrious campaigns, had wandered so far as these few miles from their respective homes, took advantage of the lull succeeding their first triumph at Berg Isel, to come over and visit the strange labyrinths of the Salzberg. It is hardly possible to exaggerate the effect which such a scene might produce on minds so imaginative, and at the same time so unsophisticated. It is not difficult to believe that they regarded such a journey like a visit to the abode of the departed great, or that in presence of the oppressive grandeurs of nature they should have matured their spirit for the defence of their country which was to confound the strategy of practised generals.

Returning through the dark forests of pine and the steep cliffs of the Hallthal, otherwise called the Salzthal, you are arrested by the hamlet of Absam, which in your hurry to push forward you overlooked in the morning. Before reaching it you observe to the east, on an eminence rising out of the plain, Schloss Melans, now serving as a villa to a family of Innsbruck. The peasants have a curious story to account for the rudely sculptured dragons which adorn some of the eave-boards of their houses, though no singular mode of ornamentation, and by others accounted for differently.[1] They say that in olden time there was a wonderful old hen which laid her first egg when seven years old, and when the egg was hatched a dragon crept out of it,[2] which made itself a home in the neighbouring moor, and the people in memory of the prodigy carved its likeness on their houses.

In Absam itself once lived a noble family of the name of Spaur, which had a toad for a bearing on their shield, accounted for in the following way:—'A certain Count Spaur had committed a crime by which he had incurred the penalty of death; his kinsmen having put every means in motion to get the sentence remitted,

---

[1] Colin de Plancy, *Légendes des sept péchés capitaux*, Appendice; and Nork, *Mythologie der Volkssagen*, point out that the dragon, sacred to Wodin, was placed on houses, town gates, and belfries, as a talisman against evil influences. See also some remarks on the two-fold character of dragons in mythology in Cox's *Mythology of the Aryan Nations*, i. 428.

[2] Compare Leoprechting, *Aus dem Lechrain*, page 78. Müllenhof *Sagen der Herzogthümer Schleswig Holstein u. Lauenburg*, page 237.

his pardon was at last accorded them on the condition that he should ride to Babylon the Accursed, and bring home with him a monstrous toad which infested the tower. So the knight rode forth to Babylon the Accursed, and when he drew near the tower the monstrous toad came out and seized the bridle of the knight's horse; the knight, nothing daunted at the horrid apparition, lifted his good sword and hewed the monster to the ground, bringing the corpse back with him as a trophy.'

What audacious tales! Could anyone out of a dream put such ideas together? No writer of fiction, none but one who believed them possible of accomplishment! 'Who can tell what gives to these simple old stories their irresistible witchery?' says Max Müller. 'There is no plot to excite our curiosity, no gorgeous description to dazzle our eyes, no anatomy of human passion to rivet our attention. They are short and quaint, full of downright absurdities and sorry jokes. We know from the beginning how they will end. And yet we sit and read and almost cry, and we certainly chuckle, and are very sorry when

> Snip, snip, snout,
> This tale's told out.

Do they remind us of a distant home—of a happy childhood? Do they recall fantastic dreams long vanished from our horizon, hopes that have set never to rise again? . . . . Nor is it dreamland altogether. There is a kind of *real* life in these tales—life such as a child believes in—a life where good is always rewarded;

wrong always punished; where everyone, not excepting the devil, gets his due; where all is possible that we truly want, and nothing seems so wonderful that it might not happen to-morrow. We may smile at those dreams of inexhaustible possibility, but in one sense the child's world is a real world too.'

A singular event, or curious popular fancy, obtained for Absam the honour of becoming a place of pilgrimage at the end of the last century. It was on January 17, 1797; a peasant's daughter was looking idly out of window along the way her father would come home from the field; suddenly, in the firelight playing on one of the panes, she discerned a well-defined image of the Blessed Virgin, 'as plain as ever she had seen a painting.' Of course the neighbours flocked in to see the sight, and from them the news of the wonderful image spread through all the country round; at last it made so much noise, that the Dean of Innsbruck resolved to investigate the matter. A commission was appointed for this object, among their number being two professors of chemistry, and the painter, Joseph Schöpf. Their verdict was that the image had originally been painted on the glass; that the colours, faded by time, had been restored to the extent then apparent by the action of the particular atmosphere to which they had been exposed. The people could not appreciate their arguments, nor realize that any natural means could have produced so extraordinary a result. For them, it was a miraculous image still, and accordingly they put their faith in it as such; nor was their faith without its fruit. It was a season of terrible trouble, a pestilence was raging both among men and beasts; General Joubert had penetrated as far into the interior as Sterzing; everyone felt the impotence of 'the arm of flesh' in presence of such dire calamities. The image on the peasant's humble window-pane seemed to have come as a token of heavenly favour; nothing would satisfy them but that it should be placed on one of the altars of the church, and the '*Gnadenmutter*[1] *von Absam*' drew all the fearful and sorrowing to put their trust in Heaven alone. Suddenly after this the enemy withdrew his troops, the pestilence ceased its havoc, and more firmly than ever the villagers believed in the supernatural nature of the image on the window-pane.

Absam has another claim to eminence in its famous violin-maker, Jacob Stainer, born in 1649. He learnt his art in Venice and Cremona, and carried it to such perfection, that his instruments fetched as high prices as those made in Cremona itself. Archduke Ferdinand Karl, *Landesfürst* of Tirol, attached him to his court. Stainer was so particular about the wood he used, that he always went over to the Gletscher forest clearings to select it, being guided in the choice by the sound it returned when he struck it with a hammer. Towards the end of his life the excitement of the love of his calling overpowered his strength of mind, and

[1] Mother of mercy.

the treatment of insanity not being then brought to perfection at Absam, one has yet to go through the melancholy exhibition of the stout oaken bench to which he had to be strapped or chained when violent.[1]

Mils affords the object of another pleasant excursion from Hall, reached through the North, or so called Mils, gate, in an easy half-hour; around it are the old castles of Grünegg and Schneeburg, the former a hunting-seat of Ferdinand II., now in ruins; the latter well-preserved by the present noble family of the name. Those who have a mind to enjoy a longer walk, may hence also find a way into the peaceful shady haunts of the Gnadenwald. Some two hundred years ago there lived about half way between Hall and Mils a bell-founder, who enjoyed the reputation of being a very worthy upright man, as well as one given to unfeigned hospitality: so that not only the weather-bound traveller, but every wayfarer who loved an hour's pleasant chat, knocked, as he passed by, at the door of the *Glockenhof*. Among all the visitors who thus sat at his board, none were so jovial as a party of wild fellows, whose business he was never well able to make out. They always brought their own meat and drink with them, and it was always of the best; and money seemed to them a matter of no account, so abundant was it. At last he ventured one day to inquire whence they acquired their seemingly bound-

[1] A touching story has been made out of his history in *Alpen Blumen Tirols*.

M

less wealth. 'Nothing easier, and you may be as rich as we, if you will!' was the answer; and then they detailed their exploits, which proved them knights of the road. Opportunity makes the thief. The proverb was realized to the letter; the *Glockengiesser* had been honest hitherto, because he had never been tempted before; now the glittering prize was exposed to him, he knew not how to resist. His character for hospitality made the *Glockenhof* serve as a very trap. The facility increased his greed, and his cellars were filled with spoil and with the skeletons of the spoiled. Travellers thus disappeared so frequently that consternation was raised again and again, but who could ever suspect the worthy hearty *Glockengiesser!* Though the new trade throve so well, there was one quality necessary to its success in which the *Glockengiesser* was wanting, and that was caution. Just as if there had been nothing to hide, he let a party of sewing-women come one day from the village to set his household goods in order; and when they retired to rest at night, one of them, who could not sleep in a strange house, heard the master and his gang counting their money in the cellar. Astonished, she crept nearer, and overheard their talk. 'We should not have killed that fellow,' said one; 'he wasn't worth powder and shot.' 'Pooh!' replied another, 'you can't expect to have good luck out of *every* murder. Why, how often a cattle-dealer kills a beast and doesn't turn a penny out of it.' The seamstress did not want to hear any more;

she laid her charge at the town-hall of Hall next morning; the officers of justice arrested the bell-founder and his associates, and ample proofs of their guilt were found on the premises. Sentenced to death, in the solitude of his cell, he yielded to the full force of the reproaches of his conscience; he made no defence, but hailed his execution as a satisfaction of which his penitent soul acknowledged the justice. However, he craved two favours before his end; the one, to be allowed to go home and found a bell for the *lieb' Frau Kirche* in Mils; the second, that this bell might be sounded for the first time at his execution, which by local custom must be on a Friday evening at nine o'clock.[1] Both requests were granted, and his bell continued to serve the church of Mils till the fire of August 1791.

Another walk from Hall is the Loreto-Kirche, intended as an exact copy of 'the Holy House,' by Archduke Ferdinand and his wife, the pious Anna Katharina of Gonzaga, who endowed it with a foundation for perpetual Masses for the repose of the souls of the reigning House of Austria; it was at one time a much visited pilgrimage, so that though it had three chaplains attached to it, monks from Hall had often to be sent for to supplement their ministrations. Ferdinand and Anna often made the pilgrimage on foot from Innsbruck, saying the 'stations' as they went, at certain little

---

[1] This was designed so as to coincide with the time when the faithful throughout the world were saying the *De Profundis*.

chapels which marked them by the way, and of which remains are still standing. It would be an interesting spot to trace out: I regret that we neglected to do so, and I do not know whether it is now well kept up.

Starting again by the North gate of Hall, and taking the way which runs in the opposite direction from that leading to Mils, you come, after half an hour's walk through the pleasant meadows, to Heiligen Kreuz; its name was originally Gumpass, but it had its present name from the circumstance of a cross having been carried down stream by the Inn, and recovered from its waters by some peasants from this place, by whom it was set up here. So great is the popular veneration for any even apparent act of homage of Nature to 'Nature's God,' that great crowds congregated to see the cross which had been brought to them by the river; and it was found necessary in the seventeenth century to erect the spot into a distinct parish. Heiligen Kreuz is much resorted to for its sulphur baths, also by people from Hall as a pleasing change from their smoky town, on holidays.

Striking out towards the mountains, another half-hour brings you to Taur, a charming little village, standing in the shelter of the Taureralpe. Almost close above it is the Thürl, a peak covered to a considerable height with rich pasture; at its summit, a height of 6,546 ft., is a wooden pyramid recording that it was climbed by the Emperor Francis I., and called the Kaisersaüle. There are many legends of S. Romedius connected with Taur, one of which is worth citing, in illustration of the confidence of the age which conceived or adapted it, in the efficacy of faith and obedience. S. Romedius was a rich Bavarian, who in the fourth century owned considerable property in the Innthal, including Taur. On his return from a pilgrimage to Rome, he put himself at the disposal of S. Vigilius, the apostle of South Tirol, who despatched him to the conversion of the Nonsthal, where he lived and died in the odour of sanctity. He was not unmindful of his own Taur, but frequently visited it to pour out his spiritual benedictions. He was once there on such a visit, when he received a call from S. Vigilius. Regardless of his age and infirmities, he immediately prepared for the journey over the mountains to Trent. His nag, old and worn out like his master, he had left to graze on the pastures at the foot of the Taureralpe, so he called his disciple David, and bid him bring him in and saddle him. Great was the consternation of the disciple on making the discovery that the horse had been devoured by a bear. Saddened and cast down, he came to his master with the news. Nothing daunted, S. Romedius bid him go back and saddle the bear in its stead. The neophyte durst not gainsay his master, but went out trusting in his word; the bear meekly submitted to the bidding of the holy man, who bestrode him, and rode on this singular mount into Trent. It is only a fitting sequel that the legend adds, that at his approach all the bells of Trent rang out a gladsome peal of welcome, without being moved by human hands.

The lords of Taur gave the name to the place by setting up their castle on the ruins of an old Roman tower (*turris;* altromanisch, *tour*). S. Romedius is not the only hero from among them; the chronicles of their race are full of the most romantic achievements; perhaps not the least of these was the construction of the fortress, the rambling ruins of which still attest its former greatness. Overhanging the bank of the Wildbach is the chapel of S. Romedius, inhabited by a hermit as lately as the seventeenth century, though the country-people are apt to confuse him with S. Romedius himself![1] One dark night, as he was watching in prayer, he heard the sound of tapping against his cell window. Used to the exercise of hospitality, he immediately opened to the presumed wayfarer: great was his astonishment to see standing before him the spirit of the lately deceased parish priest, who had been his very good friend. 'Have compassion on me, Frater Joshue!' he exclaimed; 'for when in the flesh I forgot to say three Masses, for which the stipend had been duly provided and received

---

[1] A similar fact for the comparative mythologist is recorded p. 123-4, in the case of the Bienerweible. While these sheets were preparing for the press, a singular one nearer home was brought under my notice. A little girl being asked at a national school examination, 'What David was before he was made king?' answered, 'Jack the Giant-killer.' This is a noteworthy instance of the hold of myths on the popular mind; it did not proceed from defective instruction, for the school is one of the very first in its reports, and the child not at all backward.

by me, and now my penance is fearful;' as he spoke he laid his hand upon a wooden tile of the hermit's lowly porch, who afterwards found that the impression of his burning hand was branded into the wood. 'Now do you, my friend, say these Masses in my stead; pray and fast for me, and help me through this dreadful pain.' The hermit promised all he wished, and kept his promise; and when a year and a day had passed, the spirit tapped again at the window, and told him he had gained his release. The tile, with the brand-mark on it, may be seen hanging in the chapel, with an inscription under it attesting the above facts, and bearing date 7th February, 1660.[1]

At a very short distance further is another interesting little village, Rum by name. It is situated close under the mountains, the soil of which is very friable. A terrible landslip occurred in 1770; the noise was heard as far as Innsbruck, where it was attributed to an earthquake. Whole fields were covered with the *débris,* some of which were said to be carried to a distance of a mile and a half; the village just escaped destruction, only an outlying smithy, which was buried, showed how near the danger had come. If time presses, this excursion may be combined with the last, and the Loreto-Kirche taken on the way back to Hall.

[1] Concerning *der feurige Mann,* and the mark of his burning hand, see Stöber *Sagen des Elsasses,* p. 222-3.

CHAPTER VI.

NORTH TIROL—UNTERINNTHAL (RIGHT INN-BANK).

SCHWATZ.

> *The world is full of poetry unwrit;*
> *Dew-woven nets that virgin hearts enthrall,*
> *Darts of glad thought through infant brains that flit,*
> *Hope and pursuit, loved bounds and fancies free—*
> *Poor were our earth of these bereft.* . . .
> AUBREY DE VERE.

IT is time now to return to speak of Schwatz, of which we caught a glimpse across the river as we left Viecht;[1] and it is one of the most interesting towns, and centres of excursions, in Tirol. It was a morning of bright promise which first brought us there by the early hour of 8.15. To achieve this we had had to rise betimes; it was near the end of August, when the mid-day sun is overpowering; yet the early mornings were very cool, and the brisk breezes came charged with a memory of snow from the beautiful chains of mountains whose base we were hugging. The railway station, as if it dared not with its modern innovation invade the rural retreat of primitive institutions, was at a considerable distance from the village, and we had a

---

[1] At page 145.

walk of some fifteen or twenty minutes before we came within reach of even a chance of breakfast.

My own strong desire to be brought quite within the influence of Tirolean traditions perhaps deadened my sensations of hunger and weariness, but it was not so with all of our party; and it was with some dismay we began to apprehend that the research of the primitive is not to be made without some serious sacrifice of '*le comfortable.*'

Our walk across the fields at last brought us to the rapid smiling river; and crowning the bridge, stood as usual S. John Nepomuk, his patient martyr's face gazing on the effigy of the crucified Saviour he is always portrayed as bearing so lovingly, seeming so sweetly all-enduring, that no light feeling of discontent could pass him unrestrained. Still the call for breakfast is an urgent one with the early traveller, and there seemed small chance of appeasing it. Near the station indeed had stood a deserted building, with the word '*Restauration*' just traceable on its mouldy walls, but we had felt no inclination to try our luck within them; and though we had now reached the village, we seemed no nearer a more appetising supply. No one had got out of the train besides ourselves; not a soul appeared by the way. A large house stood prominently on our right, which for a moment raised our hopes, but its too close proximity to a little church forbad us to expect it to be a hostelry, and a scout of our party brought the intelligence that it was a hospital; another building further on, on the left, gave promise again, because painted all over with frescoes, which might be the mode in Schwatz of displaying a hotel-sign; but no, it proved to be a forge, and like the lintels marked by Morgiana's chalk, all the houses of Schwatz—as indeed most of the houses of Tirol—were found to be covered with sacred frescoes. At last a veritable inn appeared, and right glad we were to enter its lowly portal and find rest, even though the air was scented by the mouldering furniture and neighbouring cattle-shed; though the stiff upright worm-eaten chairs made a discordant grating on the tiled floor, and a mildewed canvas, intended to keep out flies, completed the gloom which the smallness of the single window began.

A repeated knocking at last brought a buxom maid out of the cow-shed, who seemed not a little amazed at our apparition. 'Had she any coffee! coffee, at that time of day—*of course not!*' True, the unpunctuality of the train, the delivery of superfluous luggage to the care of the station-master, and our lingering by the way, had brought us to past nine o'clock—an unprecedented hour for breakfast in Schwatz. 'Couldn't we be content with wine? in a couple of hours meat would be ready, as the carters came in to dine then.' Meat and coffee at the same repast, and either at that hour, were ideas she could not at first take in. Nevertheless, when we detailed our needs, astonishment gave way to compassion, and she consented to drop her incongruous propositions, and to make us happy in our own way. Accordingly, she was soon busied in lighting a fire, running to fetch coffee and rolls—though she did not, as happened to me in Spain, ask us to advance the money for the commission—and very soon appeared with a tray full of tumblers and queer old crockery. The black beverage she at last provided consisted of a decoction of nothing nearer coffee than roasted corn, figs,[1] or acorns; and the rolls had the strangest resemblance to leather; but the milk and eggs were fine samples of dairy produce, for which Schwatz is famous, and these and the luscious fruit made up for the rest.

I remember that the poet-author of one of the most charming books of travel, in one of the most charming countries of Europe,[2] deprecates the habit travel-writers have of speaking too much about their fare; and in one sense his remarks are very just. Where this is done without purpose or art, it becomes a bore; but 'love itself can't live on flowers;' and as, however humiliating the fact, it is decreed that the only absolutely necessary business of man's life is the catering for his daily bread, it becomes interesting to the observant to study the various means by which this decree is complied with by different races, in different localities. It is especially noteworthy that it is just in countries made supercilious by their culture that

---

[1] '*Feigen-Kaffee*,' made of figs roasted and ground to powder, is sold throughout Austria.

[2] Aubrey de Vere's *Greece and Turkey*.

these matters of a lower order engross the most attention, and just those who consider themselves the most civilized who are the most dependent on what have been termed mere 'creature comforts.' These poor country folks, whom the educated traveller often passes by as unworthy of notice in their benighted ignorance and superstition—while they would not forego their salutation of the sacred symbol by the way-side, which marks their intimate appreciation of truths of the highest order—put us to shame, by their indifference to sublunary indulgences. We had come to Tirol to study their ways, and I hope we took our lesson on this occasion, well. We were not feasted with a sumptuous repast, such as might be found in any of the monster hotels, now so contrived, that you may pass through all the larger towns of Europe with such similarity to home-life everywhere, that you might as well never have left your fireside; but we were presented with an experience of the struggle with want; of that hardy face-to-face meeting with the great original law of labour, which our modern artificial life puts so completely out of sight, that it grows to regard it as an antiquated fable, and which can only be met amid such scenes.

The matutinal peasants were packing up their wares—which when spread out had made a picturesque market of the main street—by the time we again sallied forth, and we were nearly losing what is always one of the prettiest sights in a foreign town. At the end rose the parish church, with a stateliness for which the smallness of the village had not prepared us; but Schwatz has a sad and eventful history to account for the disparity.

Schwatz was once a flourishing Roman station, and even now remains are dug out which attest its ancient prosperity; but it had fallen away to the condition of a neglected *Hausergruppe* by the fourteenth century, when suddenly came the discovery of silver veins in the surrounding heights. A lively bull,[1] one day tearing up the soil with his horns in a frolic, laid bare a shining vein of ore. The name of Gertraud Kandlerin, the farm-servant who had charge of the herd to which he belonged, and brought the joyful tidings home to Schwatz, has been jealously preserved. From that moment Schwatz grew in importance and prosperity: and at one time there was a population of thirty thousand miners employed in the immediate neighbourhood. The Fuggers and Hochstetters of Augsburg were induced to come and employ their vast resources in working the riches of the mountains; and native families of note, laying aside the pursuit of arms, joined in the productive industry. Among these were the Fiegers, one of whom was the counsellor and intimate friend of the Emperor Maximilian, who followed his remains to their last resting-place, at Schwatz, when he died in ripe old age, leaving fifty-seven children and grandchildren, and money enough

[1] Burglechner. A.D. 1409.

to enrich them all. His son Hanns married a daughter of the Bavarian house of Pienzenau; and when he brought her home, tradition says it was in a carriage drawn by *four thousand* horses. Many names, famous in the subsequent history of the country, such as the Tänzls, Jöchls, Tannenbergs, and Sternbachs, were thus first raised to importance. This outpouring of riches stimulated the people throughout the country to search for mineral treasures, and everywhere the miners of Schwatz were in request as the most expert, both at excavating and engineering. Nor this only within the limits of Tirol; they had acquired such a reputation by the middle of the sixteenth century, that many distant undertakings were committed to them too. They were continually applied to, to direct mining operations in the wars against the Turks in Hungary. Their countermines performed an effective part in driving them from before Vienna in 1529; and again, in 1739, they assisted in destroying the fortifications of Belgrad. Clement VII. called them to search the mountains of the Papal State in 1542; and the Dukes of Florence and Piedmont also had recourse to their assistance about the same time. In the same way, many knotty disputes about mining rights were sent from all parts to be decided by the experience of Schwatz; and its abundance attracted to it every kind of merchandise, and every new invention. One of the earliest printing-presses was in this way set up here.

But a similarity of pursuit had established a community of interest between the miners of Schwatz and their brethren of Saxony; and when the Reformation broke out, its doctrines spread by this means among the miners of Schwatz, and led at one time to a complete revolution among them. Twice they banded together, and marched to attack the capital, with somewhat communistic demands. Ferdinand I., and Sebastian, Bishop of Brixen, went out to meet them on each occasion at Hall, and on each occasion succeeded in allaying the strife by their moderate discourse. Within the town of Schwatz, however, the innovators carried matters with a high hand, and at one time obtained possession of half the parish church, where they set up a Lutheran pulpit. Driven out of this by the rest of the population, they met in a neighbouring wood, where Joham Strauss and Christof Söll, both unfrocked monks, used to hold forth to them.

A Franciscan, Christof von München, now came to Schwatz, to strengthen the faith of the Catholics, and the controversy waged high between the partisans of both sides; so high, that one day two excited disputants carried their quarrels so far before a crowd of admiring supporters, that at last the Lutheran exclaimed, 'If Preacher Söll does not teach the true doctrine, may Satan take me up into the Steinjoch at Stans!' and as he spoke, so, says the story, it befell: the astonished people saw him carried through the air and disappear from sight! The credit of the Lutherans fell very sensibly on the instant, and still more some days after, when the adventurous victim came back lame and bruised, and himself but too well convinced of his error.

Nevertheless the strife was not cured. Somewhat later, there was an inroad of Anabaptists, under whose auspices another insurrection arose, and for the time the flourishing mining works were brought to a stand-still. At last the Government was obliged to interfere. The most noisy and perverse were made to leave the country, and the Jesuits from Hall were sent over to hold a mission, and rekindle the Catholic teaching. Peace and order were restored : four thousand persons were brought back to the frequentation of the sacraments; but the *Bergsegen*,[1] add the traditions, which had been the occasion of so much dissunion, was never recovered. From that time forth the mining treasures of Schwatz began to fail; and after a long and steadily continued diminution of produce, silver ceased altogether to be found. Copper, and the best iron of Tirol, are still got out, and their working constitutes one of the chief industries of the place ; the copper produced is particularly fit for wire-drawing, for which there is an establishment here. Another industry of Schwatz is a government cigar manufactory,[2] which employs between four and five hundred hands, chiefly women and children, who get very poorly paid—ten or twelve

[1] Mineral wealth—*lit.* Mountain-blessing.
[2] I was told there that it had been reckoned that 500,000 cigars are smoked per diem in Tirol.

francs a-week, working from five in the morning till six in the evening, with two hours' interval in the middle of the day. There are pottery works, which also employ many hands; and many of the women occupy themselves in knitting woollen clothing for the miners. The pastures of the neighbourhood are likewise a source of rich in-comings to the town ; but with all these industries together, Schwatz is far below the level of its early prosperity. Instead of its former crowded buildings, it now consists almost entirely of one street; and instead of being the cynosure of foreigners from all parts, is so little visited, that the people came to the windows to look at the unusual sight of a party of strangers as we passed by. In place of its early printing-press, its literary requirements are supplied by one little humble shop, where twine, toys, and traps, form the staple, and stationery and a small number of books are sold over and above : and where, because we spent a couple of francs, the master thereof seemed to think he had driven for that one day a roaring trade.

Other misfortunes, besides the declension of its 'Bergsegen,' have broken over Schwatz. In 1611 it was visited by the plague, in 1670 by an earthquake; but its worst disaster was in the campaign of 1809, when the Bavarians, under the Duke of Dantzic, and the French, under Deroi, determined to strike terror into the hearts of the country-people by burning down the town. The most incredible cruelties are reported to have been perpetrated on this occasion, many being such as one cannot bear to repeat; so determined was their fury, that when the still air refused to fan the flames, they again and again set fire to the place at different points; and the people were shot down when they attempted to put out the conflagration. General Wiede was quartered in the palace of Count Tannenberg, a blind old man, with four blind children; his misfortunes, and the laws of hospitality, might have protected him at least from participation in the general calamity; but no, not even the hall where the hospitable board was spread in confidence for the unscrupulous guest, was spared. Once and again, as the inimical hordes poured into, or were driven out of, Tirol, Schwatz had to bear the brunt of their devastations, so that there is little left to show what Schwatz was. The stately parish church, however, suffered less than might have been expected: in the height of the conflagration, when all was noise and excitement, a young Bavarian officer, over whom sweet home lessons of piety exercised a stronger charm than the wild instincts of the military career which were effecting such havoc around, collected a handful of trusty followers, and, unobserved by the general herd, succeeded in rescuing it before great damage had been done.

The building was commenced about 1470,[1] and consecrated in 1502. What remains of the original work is in the best style of the period; the west front

---

[1] The date of death on the tombstone of Lukas Hirtzfogel, whom tradition calls the architect of this church, is 1475.

is particularly noteworthy. The plan of the building is very remarkable, consisting of a double nave, each having its aisles, choir, and high-altar; this peculiar construction originated in the importance of the *Knappen,* or miners, at the time it was designed, and their contribution to the building fund entitling them to this distinct division of the church between them and the towns-people; one of the high-altars still goes by the name of the *Knappenhochaltar.* The roof, like those of most churches of Tirol and Bavaria, is of copper, and is said to consist of fifteen thousand tiles of that metal—an offering from the neighbouring mines. The emblem of two crossed pick-axes frequently introduced, further denotes the connexion of the mining trade with the building. Whitewash and stucco have done a good deal to hide its original beauties, but some fine monuments remain. One in brass, to Hanns Dreyling the metal-founder, date 1578, near the side ('south') door, should not be overlooked: the design embodies a Renaissance use of Ionic columns and entablature in connexion with mediæval symbols. Below, are seen Hanns Dreyling himself in the dress of his craft, his three wives, and his three sons habited as knights (showing the rise of his fortunes), all under the protection of S. John the Baptist. Above, is portrayed the vision of the Apocalypse, God the Father seated on His Throne, surrounded by a rainbow, with the Book of Seven Seals, and the Lamb; at His Feet the four Evangelists; around, the four-and-twenty elders, with their harps, some wearing their crowns, and some stretching them out as a humble offering before the Throne; in front kneels the Apostolic Seer himself, gazing, and with his right hand pointing, upwards, yet smiting his breast with the left hand, and weeping that no one was found worthy to open the seals of the book. Below the epitaph, the monument bears the following lines :

> Mir gab Alexander Colin den Possen
> Hanns Löffler hat mich gegossen.

Alexander Colin, of Malines, and Hans Löffler, were, like Hans Dreyling, *Schmelzherrn* of eminence, and connected with him by marriage, thus they naturally devoted their best talent to honour their friend and master. We learnt to appreciate it better when we came to see their works at Innsbruck. The nine altar-pieces are mostly by Tirolean painters. The Assumption, on one high-altar, is by Schöpf; the Last Supper on the other—the *Knappenhochaltar*—by Bauer of Augsburg.

The 'north' side door opens on to a narrow strip of grass, across which is a *Michaels-kapelle*, as the chapel we so often find in German churchyards—and where the people love to gather, and pray for their loved and lost—is here called. It is a most beautiful little specimen of middle-pointed, with high-pitched roof and traceried window. A picturesque stone-arched covered exterior staircase, the banister cornice of which represents a narrow water-trough, with efts chasing each other in and out of it, leads to the upper chapel, which was in some little confusion at the time of our visit, as it was under restoration; two or three artists were in the lower chapel, painting the images of the saints in the fresh colours the people love. After some searching, I found out a figure of a dead Christ, which I was curious to see; because, before coming to Schwatz, I had been told there was one which had been dearly prized for centuries by the people; that once on a time there had come night by night a large toad, and had stood before the image, resting on its hinder feet, the two front ones joined as if in token of prayer; and no one durst disturb it, because they said it must be a suffering soul which they saw under its form. I spoke to one of the artists about it, to see if this was the right image, and if the legend was still acknowledged. He answered as one who had little sympathy with the mysteries he was employed to delineate; he evidently cared nothing for legends, though willing to paint them for money. It was the first time I had met with this sort of spirit in the neighbourhood, and was not surprised to learn he was not of Tirol, but from Munich.

A door opposite the last named opens into the churchyard, filled with the usual black and gold cast-iron crosses, and the usual sprinkling of some of a brighter colour; each with its stoup of holy water and *weihwedel*,[1] and its simple epitaph, 'Hier ruhet in

---

[1] Brush for sprinkling holy-water.

Friede. . . . .' Besides the large crucifix, which always stands in the centre promising redemption to the faithful departed, is a stout round pillar of large rough stones, surmounted by a lantern cap with five sharp points, each face glazed, and a lamp within before some relic, always kept alight, for the people think[1] that the holy souls come and anoint their burning wounds with the oil which piously feeds a churchyard lamp. Twinkling fitfully amid the evening shadows, over the graves, and over the human skulls and bones, of which there happened fortuitously to be a heap waiting re-sepulture after some late arrangement of the burying-ground, it disposed one to listen to the strange tales which are told of it. There was once a *Robler* of Schwatz, well-limbed, deep-chested, full of confidence and energy, who had won the right to wear the champion feather[2] against the whole neighbourhood. But not content to be the darling of his home, and the pride of his valley, he must needs prove himself the best against all comers. In fear of the shame of a reverse after all his boasts, he resolved to ensure himself against one, by having recourse to an act, originally designed probably as a test of possessing, but commonly believed by the people to be a means of winning, invincible strength of nerve, and which is described in the following narrative. Opportunity was not lacking. Death is ever busy, and one day laid low an old gossip, who was duly buried

---

[1] See note to p. 140.
[2] See p. 95.

with all honour by her children and children's children to the third generation. Now was the time for our brave *Robler*. That first night that she rested in the 'field of peace,' he rose in the dead of the night—a dark starless night, just as it was when we stood there—and the lamp of the shrine resting its calm pale rays upon the graves. The great clock struck out twelve, with a rattling of its cumbersome machinery, which sounded like skeletons walking by in procession: our *Robler* quailed not, however, but approached the new grave, scattered the earth from over it with his spade, raised up the coffin, opened it, took out the corpse, dressed himself in its shroud, and lifted the ghastly burden on to his strong shoulders. Never had burden felt so heavy; it seemed to him as though he bore the Freundsberg on his back; though sinking and quailing, he bore it three times round the whole circuit of the enclosure, laid it back in the coffin, and lowered the coffin into the grave; triumphantly he showered the earth over it, and took quite a pleasure in shaping the hillock smoothly and well. Then suddenly, to his horror, with a click like the gripe of a skeleton, he heard the clapper of the old clock raised to mark the completion of the hour within which his task, to be effectual, must be accomplished. Meantime, it had come on to rain violently, and the big drops pattered on the stones, like dead men tramping all around him; it happened to fall heavily round us, and the simile was so striking, I could not forbear a grim smile. It seemed to him as if he never could dash through their midst in time; still he made the attempt boldly, and actually succeeded in swinging himself over the churchyard wall before the hammer had fallen, and, what was most important, still bearing round his shoulders the shroud of the dead. Nevertheless his heart was full of anxiety with the thought that he had disturbed the peace of the departed; it seemed to him as if the old gossip had run after him to claim her own, and with her burning hand had seized the fluttering garment, and torn a piece out of it, just as he cleared the wall. For days after, the sexton saw the piece, torn and burnt, fluttering over her grave, but never could make out how it got there. The *Robler*, however, was now proof against every attack; no one could wear a feather in his presence, for he was sure to overcome him, and make him renounce the prize. What did he gain, however, by his uncannily-earned prowess? A little temporary renown and honour, and the fear of his kind; but all through the rest of his life, at the *Wandlung*[1] of the Holy Mass, the pure white wafer, as the priest raised it aloft, seemed *black* to his eyes, and when he came to die, there was no father-confessor near to whisper absolution and peace.

A most singular legend, also attached to this spot, dates from the time when the Jesuit Fathers held their missions after the expulsion of the Lutherans.

---

[1] See note to p. 48.

With the fervour of new conversion, the people ascribed to their word the most wonderful powers; and their simple unwavering faith seems to have been a loan of that which removes mountains. Among those whom a spirit of penance moved to come and make a general confession of their past lives was a lady no longer young, of blameless character, but unmarried. The fathers, as I have already implied, enjoyed the most unbounded confidence of the people; and the most unusual penance was accepted in the simplest way. To this person the penance enjoined was, that she should for three nights watch through the hour of midnight in the church, and then come and give an account of what she had seen. Being apparently a person of a strong mind, she was satisfied with the assurance of the father that no harm would happen to her, and she fulfilled her task bravely. When she came to narrate what had passed, she said that each night the church had been traversed by a countless train of men, women, and children, of every age and degree, dressed in a manner unlike anything she had seen or read of in the past; the features of all quite unknown to her, and yet exhibiting a certain likeness, which might lead her to believe they might be of her own family, and all wearing an expression indescribably sad; she was all anxiety to know what she could do for their relief, for she felt sure it was to move her to this that they had been revealed to her. The father told her, however, this was not at all the object of the vision: that the train of people she had seen was an appearance of the generations of unborn souls, who might have lived to the eternal honour and praise of God if she had not preferred her ease and freedom and independence to the trammels of the married state; 'for,' said he, 'your choice of condition was based on this, not on the higher love of God, and the desire of greater perfection. Now, therefore, reflect what profit your past life has borne to the glory of God, and strive to make it glorify Him in some way in the future.'

The Franciscan church was built about the same time as the other, and has some remains of the beautiful architecture of its date. Over the credence table is a remarkable and very early painting on panel, of the genealogy of our Lord. Within the precincts of the monastery are some early frescoes, which I did not see; but they ought not to be overlooked. One subject, said to be very boldly and strikingly handled, is the commission to the Apostles to go out and preach the Gospel to the nations.

The day was wearing on, and we had our night's lodging to provide; the inn where we had breakfasted did not invite our confidence, despite of the pretty Kranach's Madonna which smiled over the parlour, and the good-natured maid who deemed it her business to wait behind our chairs while we sipped our coffee; so we walked down the long street, and tried our luck at one and another. There were plenty of them: and they were easily recognized now we knew their token, for each has a forbidden-fruit-tree painted on the wall with some subject out of the New Testament surmounting it, to show the triumph of the Gospel over the Fall; while the good gifts of Providence, which mine host within is so ready to dispense, are typified by festoons of grape-vines, surrounding the picture. Those which let out horses have also a team cut out in a thin plate of copper, and painted *proper*, as heralds say, fixed at right angles to the doorpost. Nevertheless, the interiors were not inviting, and at more than one the bedding was all on the roof, airing; and the solitary maid, left in charge of the house while all the rest of the household were in the fields harvesting, declared the impossibility of getting so many beds as we wanted ready by the evening. Dinner at the *Post* having somehow indisposed us for it, we at last put up at the *Krone*, which was very much like a counterpart of our first experience. Nothing could exceed the pleasant willingness of the people of the house; but both their accommodation and their cleanliness was limited; and besides a repulsive look, there was an unaccountable odour, about the beds, which made sleeping in them impossible. My astonishment may be imagined, when on proceeding to examine whether there were any articles of bedding that would do to roll oneself up in on the floor, I found that the smell proceeded from layers of apples between the mattresses, which it seems to be the habit thus to preserve for winter use!

The rooms were large and rambling, and filled with

cumbersome furniture, some of which must, I think, have been made before the great fire of 1809. As in all the other houses, a guitar hung on the wall of the sitting-room; and after many coy refusals, the daughter of the house consented to sing to it one or two melodies very modestly and well.

You do not sleep very soundly on the floor, and by six next morning the tingling of the Blessed Sacrament bell sufficed to rouse me in time to see how the Schwatzers honour '*das hochwürdigste Gut*,'[1] as it passed them on its way to the sick. Two little boys in red cassocks went first, bearing red banners and holywater; two followed in red and yellow, bearing a canopy over the priest, and four men carried lanterns on long poles. The rain of the previous night had filled the road with puddles, but along the whole way the peasants were on their knees. To all who are afflicted with long illnesses, it is thus carried at least every month.

The morning was bright and hot, but the ruined castle on the neighbouring Freundsberg looked temptingly near; and we easily found a rough but not difficult path, past a number of crazy cottages, the inhabitants of which, however poor and hard worked, yet gave us the cheerful Christian greeting, 'Gelobt sei Jesus Christus!' as we passed. Near the summit the cottages cease; and after a short stretch in the

---

[1] 'The most precious good,' or 'possession;' a Tirolean expression for the Blessed Sacrament.

burning sun, you appreciate the shade afforded by a tiny chapel, at the side of a crystal spring, welling up out of the ground, its waters cleverly guided into a conduit, formed of a hollowed tree, which supplies all the houses of the hill-side, and perhaps accounts for their being so thickly clustered there. The last wind of the ascent is the steepest and most slippery. The sun beat down relentlessly, but seemed to give unfailing delight to myriads of lizards, adders, and grasshoppers, who were darting and whirring over the crumbling stones in the maddest way. Historians, poets, or painters, have made some ruins so familiarly a part of the world's life, and their grand memories of departed glories have been so often recounted, that they seem stereotyped upon them. Time has shattered and dismantled them, but has robbed them of nothing, for their glories of all ages are concrete around them still. But poor Freundsberg! who thinks of it? or of the thousand and one ruined castles which mark the 'sky-line' of Tirol with melancholy beauty? Each has, however, had its throb of hope and daring, and its day of triumph and mastery, often noble, sometimes—not so often as elsewhere—base. Freundsberg is no exception. For two hundred years before the Christian era it was a fortress, we know: for how long before that we know not; and then again, we know little of what befell it, till many hundred years after, in the twelfth, thirteenth, and fourteenth centuries, its lords were known as mighty men of war. It reached its highest glory under Captain Georg, son of Ulrich and a Swabian heiress whose vast dowry tended to raise the lustre of the house.

Georg von Freundsberg entered the career of arms in early youth, and rose to be a general at an age when other men are making their *premières armes*. At four-and-twenty he was reckoned by Charles Quint his most efficient leader. Over the Swiss, over the Venetians, wherever he led, he was victorious. The victory at Pavia was in great measure due to his prowess. His personal strength is recorded in fabulous terms; his foresight in providing for his men, and his art of governing and attaching them, were so remarkable, that they called him their father, and he could do with them whatever he would. They recorded his deeds in the terms in which men speak of a hero: they said that the strongest man might stand up against him with all his energy, and yet with the little finger of his left hand he could throw him down; that no matter at what fiery pace a horse might be running away, if he but stretched his hand across the path he brought it to a stand; that in all the Emperor's stores there was no field-piece so heavy but he could move it with ease with one hand. They sang of him:

> Georg von Freundsberg,
> von grosser Sterk,
> ein theurer Held;
> behielt das Feld

> in Streit und Krieg.
> den Feind niederslieg
> in aller Schlacht.
> er legt Got zu die Er und Macht.'[1]

The last line would show that to a certain extent he was not untrue to the traditions of his country; nevertheless, his success in war, and his love for the Emperor, carried him so far away from them, that when the siege of Rome was propounded, he not only accepted a command in the attack on the 'Eternal City,' but raised twelve thousand men in his Swabian and Tirolean possessions to support the charge. None who have pondered the havoc and the horrors of that wanton and sacrilegious siege will care to extenuate the guilt of any participator in it. It is the blot on Georg von Freundsberg's character, and it was likewise his last feat. He died suddenly within the twelvemonth, aged only fifty-two, leaving his affairs in inextricable confusion, and his estate encumbered with debts incurred in raising the troops who were to assist in the desolation of the 'Holy City.'

His brothers—Ulrich, Bishop of Trent, and Thomas, who like himself followed the military calling—earned a certain share of respect also; but no subsequent member of the family was distinguished, and the race came to an end in 1580. The castle fell into ruin; and

---

[1] George of Freundsberg; a man of great strength; a worthy hero; master of the field in combat and war; in every battle the enemy fell before him. The honour and power he ascribed to God.

as if a curse rested on it, when it was used again, it was to afford cover to the Bavarians in *firing upon the people* in 1809! I do not know by what local tradition, but some motive of affection still renders the chapel a place of pious resort; and a copy of Kranach's 'Mariähilf' adorns the altar. The remaining tower affords a pleasing outline.

I returned to the chapel by the brook, and sat down to sketch it, though rather too closely placed under it to view it properly; there is always an indefinable satisfaction in making use of these places of pious rest, which brotherly charity has provided for the unknown wayfarer. When, after a time, I looked up from my paper, I saw sitting outside in the sun a strange old woman, the stealthy approach of whose shoeless feet I had not noticed. I advised her to come in and rest; and then I asked her how she came to walk unshod over the stones of the path, which were sharp and loose, as well as burning hot, while she carried a pair of stout shoes in her hand. '*That* doesn't hurt,' she replied indignantly; 'it's the shoes that hurt. When you put your foot down you know where you put it, and you take hold of the ground; but when you have those things on, you don't know where your foot goes, and down you go yourself. That's what happened to me on this very path, and see what came of it.' And she bared her right arm, and showed that it had been broken, and badly set, and now was withered and useless—she could do no more work to support herself. I asked her how she lived, and she did not like the question, for begging, it seems, is forbidden. But I said it was a very hard law, and then she grew more confidential; and after a little more talk, her wild weird style, and her strong desire to tell my fortune, showed me she was one of those dangerous devotees who may be considered the camp-followers of the Christian army, whose chance of ingratiating themselves seems greatest where the faith is brightest, and who there work all manner of mischief, overlaying simple belief with pagan superstition; but at the same time, such an one is generally a very mine for the comparative mythologist, and in this individual instance not without some excuse in her misfortunes. For, besides the unlucky disablement already named, she had lost not only her house, home, and belongings, but all her relations also, in a fire. It is not surprising if so much misery had unhinged her mind. Her best means of occupation seemed to be, when good people gave her alms, to go to a favourite shrine, and pray for them; and I fully believe, from her manner, that she conscientiously fulfilled such commissions, for I did not discover anything of the hypocrite about her. Only once, when I had been explaining what a long way I had come on purpose to see the shrines of her country, she amused me by answering, in the most inflated style, that however far it might be, it could not be so far as *she* had come—she came from beyond mountains and seas, far, far, ever so far—till I looked at her again, and wondered if she were a gipsy, and was appropriating to her personal experience some of the traditional wanderings of her race. Presently she acknowledged that her birth-place was Seefeld, which I knew to be at no great distance from Innsbruck, perhaps ten miles from where we stood. Yet this tone of exaggeration may have arisen from an incapacity to take in the idea of a greater distance than she knew of previously, rather than from any intention to deceive; and her 'seas' were of course lakes, which when spoken of in the German plural have not even the gender to distinguish them.

When she had once mentioned Seefeld, she grew quite excited, and told me no place I had come from could boast of such a marvellous favour as God had manifested to her Seefeld. I asked her to tell me about it. 'What! don't you know about Oswald Milser?' and I saw my want of recognition consigned me to the regions of her profoundest contempt. 'Don't you know about Oswald Milser, who by his pride quenched all the benefit of his piety and his liberality to the Church? who, when he went to make his Easter Communion one *grüne Donnerstag*,[1] insisted that it should be given him in one of the large Hosts, which the priest uses, and so distinguish him from the people. And when the priest, afraid to offend the great man, complied, how the weight bore him down,

[1] Maundy Thursday.

down into the earth;' and she described a circle with her finger on the ground, and bowed herself together to represent the action; 'and he clung to the altar steps, but they gave way like wax; and he sank lower and lower,[1] till he called to the priest to take the fearful Host back from him.' 'And what became of him?' I asked. 'He went into the monastery of Stamms, and lived a life of penance. But his lady was worse than he: when they told her what had taken place, she swore she would not believe it; " As well might you tell me," she said, and stamped her foot, " that that withered stalk could produce a rose;" and even as she spoke, three sweet roses burst forth from the dry branch, which had been dead all the winter. Then the proud lady, refusing to yield to the prodigy, rushed out of the house raving mad, and was never seen there again; but by night you may yet hear her wailing over the mountains, for there is no rest for *her*.' Her declamation and action accompanying every detail was consummate.

I asked her if she knew no such stories of the neighbourhood of Schwatz. She thought for a moment, and then assuming her excited manner once more, she pointed to a neighbouring eminence. 'There was a bird-catcher,' she said, 'who used to go out on the *Goaslahn* there, following his birds; but he was quite

---

[1] Stöber *Sagen des Elsasses* records a legend of a similar judgment befalling a man who, in fury at a long drought, shot off three arrows against heaven.

mad about his sport, and could not let it alone, feast day or working day. One Sunday came, and he could not wait to hear the holy Mass. "I'll go out for an hour or two," he said; "there'll be time for that yet." So he went wandering through the woods, following his sport, and the hours flew away as fast as the birds; hour by hour the church bell rang, but he always said to himself he should be in time to catch the Mass of the next hour. The nine o'clock Mass was past, and the clock had warned him that it was a quarter to ten, and he had little more than time to reach the last Mass of the day. Just as he was hesitating to pack up his tackle, a beautiful bird, such as he had never seen before, with a gay red head, came hopping close to his decoy birds. It was not to be resisted. The bird-catcher could not take his eye off the bird. "Dong!" went the bell; hop! went the bird. Which should he follow? The bird was so *very* near the lime now; there must be time to secure him, and yet reach the church, at least before the Gospel. At last, the final stroke of the bell sounded; and at the same instant the beautiful bird hopped on to the snare. Who could throw away so fair a chance? Then the glorious plumage must be carefully cleansed of the bird-lime, which had assisted the capture, and the prize secured, and carefully stowed away at home. It would be too late for Mass then; and the bird-catcher felt the full reproach of the course he was tempted to pursue, nevertheless he could not resist it. On he went, homewards; now full of buoyant joy over his luck, now cast down with shame and sorrow over his neglected duty. He had thus proceeded a good part of his way, before he perceived that his burden was getting heavier and heavier; at last he could hardly get along under it. So he set it down, and began to examine into the cause. He found that the strange bird had swelled out so big, that it was near bursting the bars of its cage, while from its wings issued furious sulphurous fumes. Then he saw how he had been deceived; that the delusive form had been sent by the Evil One, to induce him to disobey the command of the Church. Without hesitation he flung the cursed thing from him, and watched it, by its trail of lurid flame, rolling down the side of the Goaslahn. But never, from that day forward, did he again venture to ply his trade on a holy day.

'Such things had happened to others also,' she said. 'Hunters had been similarly led astray after strange chamois; for the power of evil had many a snare for the weak. Birds too, though we deemed them so pretty and innocent, were, more often than we thought, the instruments of malice.' And it struck me as she spoke, that there were more crabbed stories of evil boding in her repertory than gentle and holy ones. 'There is the swallow,' she instanced: 'why do swallows always hover over nasty dirty marshy places? Don't you know that when the Saviour was hanging on the Cross, and the earth trembled, and the sun grew dark with horror, and all the beasts of the field went and hid themselves for shame, only the frivolous[1] swallows flitted about under the very shadow of the holy rood, and twittered their love songs as on any ordinary day. Then the Saviour turned His head and reproached the thoughtless birds; and mark my words, never will you see a swallow perched upon anything green and fresh.'

I was sorry to part from her and her legendary store; but I was already due at the station, to meet friends by the train. She took my alms with glee, and then pursued her upward way barefooted, to make some promised orisons at the Freundsberg shrine.

It was a glowing afternoon; and after crossing the unshaded bridge and meadows, to and from the railway, I was glad to stop and rest in a little church which stood open, near the river. It was a plain whitewashed edifice, ornamented with more devotion than taste. When I turned to come away, I found that the west wall was perforated with a screen of open iron-work, on the other side of which was an airy hospital ward. The patients could by this means beguile their weary hours with thoughts congenial to them suggested by the Tabernacle and the Crucifix. A curtain hung by the side, which could be drawn across the screen at pleasure. There were not more than four or five patients in the ward at the time, and in most instances decay of nature was the cause of disease. There is not much illness at Schwatz; but admittance

[1] *Leichtsinnig.*

to the simple accommodation of the hospital is easily conceded. Schwatz formerly had two, but the larger was burnt down in 1809. The remaining one seems amply sufficient for the needs of the place.

There was 'Benediction' in the church in the evening, for it was, I forget what, saint's day. The church was very full, and the people said the Rosary in common before the Office began. A great number of the girls from the tobacco factory came in as they left work, and the singing was unusually sweet, which surprised me, as the Schwatzers are noted for their nasal twang and drawling accent in speaking. I learnt that there are several Italians from Wälsch-Tirol settled here, and they lead the choir. It is edifying to see the work-people, after their day's toil, coming into the church as if it was more familiar to them even than home; but one does not get used to seeing the uncovered heads of the women, though indeed with the rich and luxuriant braids of hair with which Nature endows them, they might be deemed 'covered' enough.

A more familiar sight to an English eye is the seat-filled area of the German churches. Confessedly it is one of the home associations which one least cares to see reproduced, but the pews of the German churches are less objectionable than our own; they are lower, and not so crowded, and ample space is always left for processions, so they interfere far less with the architectural design.

CHAPTER VII.

NORTH TIROL—UNTERINNTHAL (RIGHT INN-BANK).

EXCURSIONS FROM SCHWATZ.

' *Partout où touche votre regard vous rencontrez au fond sous la forme qui passe, un mystère qui demeure . . . chacun des mystères de ce monde est la figure, l'image de celui du monde supérieur ; de sorte que tout ce que nous pouvons connaître dans l'ordre de la nature est la révélation même de l'ordre divin.*'—CHEVÉ, *Visions de l'Avenir*.

FALKENSTEIN, which may be reached by a short walk from Schwatz, is worth visiting on account of the information it affords as to the mode of working adopted in the old mines both of silver and copper. This was the locality where the greatest quantity of silver was got ; it was particularly noted also for the abundance and beauty of the malachite, found in great variety and richness of tints ; the turquoise was found also, but more rarely. The old shaft runs first horizontally for some two miles, and then sinks in two shafts to a depth of some two hundred and thirty fathoms. The engineering and hydraulic works seem to have been very ingenious, but the description of them does not come within the sphere of my present undertaking. It does, however, to observe that over this, as over everything else in Tirol, religion shed its halo.

The miners had ejaculatory prayers, which it was their custom to utter as they passed in and out of their place of subterranean toil; and an appropriate petition for every danger, whether from fire-damp, land-slips, defective machinery, or other cause. Their greeting to each other, and to those they met by the way, in place of the national 'Gelobt sei Jesus Christus,' was 'Gott gebe euch Glück und Segen!'[1] For their particular patron they selected the Prophet Daniel, whose preservation in the rocky den of the lions, as they had seen it portrayed, seemed to bear some analogy with their own condition. Of their liberality in church-building I have already spoken; but many are the churches and chapels that bear the token—a crossed chisel and hammer on a red field—of their contribution to its expense.

There are many other walks to be made from Schwatz. First there is Buch, so called from the number of beech-trees in the neighbourhood, which afford pleasant shade, and diversify the scenery, in which the castle of Tratzberg across the Inn[2] also holds an important part. Further on is Margareth, surrounded by rich pastures, which are watered by the foaming Margarethenbach. Then to the south-east is Galzein, with a number of dependent 'groups of houses,' particularly Kugelmoos, the view from which sweeps the Inn from Kufstein to Innbruck. Beyond, again, but further south, is the Schwaderalpe, whence the iron worked and taken in

---

[1] God prosper and bless you!   [2] *Supra*, pp. 80-2.

depôt at Schwatz is got; and the Kellerspitze, with the little village of Troi, its twelve houses perched as if by supernatural handiwork on the spur of a rock, and once nearly as prolific as Falkenstein in its yield of silver. The exhausted—deaf (*taub*) as it is expressively qualified in German—borings of S. Anthony and S. Blaze are still sometimes explored by pedestrians.

Arzburg also is within an hour's walk. It was once rich in copper ore, but is now comparatively little worked. Above it is the *Heiligenkreuzkapelle*, about which it is told, that when, on occasion of the *baierische-Rumpel*[1] in 1703, the bridge of Zirl was destroyed, the cross which surmounted it being carried away by the current, was here rescued and set up by the country-people, who still honour it by frequent pilgrimages.

Starting again from Schwatz by the high-road, which follows pretty nearly the course of the Inn, you pass a succession of small towns, each of which heads a valley, to which it gives its name, receiving it first from the torrent which through each pours the aggregate of the mountain streams into the river, all affording a foot-way through the Duxerthal into the further extremity of the Zillerthal—Pill, Weer, Kolsass, Wattens, and Volders.

First, there is Pill, a frequent name in Tirol, and derived by Weber from *Bühl* or *Büchl*, a knoll; it is the wildest and most enclosed of any of these lateral valleys,

[1] Rout of the Bavarians.

and exposed to the ravages of the torrent, which often in winter carries away both bridges and paths, and makes its recesses inaccessible even to the hardy herdsmen. The following story may serve to show *how* hardy they are:—Three sons of a peasant, whose wealth consisted in his grazing rights over a certain tract of the neighbouring slopes, were engaged one day in gathering herbage along the steep bank for the kids of their father's flock. The steep must have been difficult indeed on which they were afraid to trust mountain kids to cater for themselves; and the youngest of the boys was but six, the eldest only fifteen. The eldest lost his balance, and was precipitated into the roaring torrent, just then swollen to unusual proportions; he managed to cling fast behind one of the rocky projections which mark its bed, but his strength was utterly unable to bear him out of the stream. The second brother, aged ten, without hesitating, embraced the risk of almost certain death, let himself down the side of the precipice by clinging to the scanty roots which garnished its almost perpendicular side. Arrived at the bottom, he sprang with the lightness of a chamois across the foaming waters on to the rock where the boy was now slackening his exhausted hold, and succeeded in dragging him up on to the surface; but even there there seemed no chance of help, far out of sound as they were of all human ears. But the youngest, meantime, with a thoughtfulness beyond his years, had made his way home alone, and apprised the father, who readily found the means of rescuing his offspring.

The break into the Weerthal is at some little distance from the high road; its church, situated on a little high-level plain, is surrounded with fir-trees. A little lake is pointed out, of which a similar legend is told to 'the judgment of Achensee,' which is indeed one not infrequently met with; it is said that it covers a spot where stood a mighty castle, once submerged for the haughtiness of its inhabitants, and the waters placed there that no one might again build on the site for ever. The greatest ornament of the valley is the rambling ruin of Schloss Rettenberg, on its woody height, once a fortress of the Rottenburgers; afterwards it passed to Florian Waldauf, whose history I have already given when speaking of Hall.[1] It was bought by the commune in 1810, and the present church built up out of the materials it afforded, the former church having been burnt down that year. The old site and its remains are looked upon by the people as haunted by a steward of the castle and his wife, who in the days of its prosperity dealt hardly with the widow and the orphan, and must now wander sighing and breathing death on all who come within their baleful influence. A shepherd once fell asleep in the noontide heat, while his sheep were browsing on the grass-grown eminence. When he woke, they were no longer in sight; at last he found them dead within the castle keep. 'Guard thy flock better,' shouted a hoarse voice, 'for this enclosure is

---

[1] See pp. 151-2.

mine, and none who come hither escape me.' None ventured within the precincts after this; but many a time those who were bold enough to peep through a fissure in the crazy walls reported that they had seen the hard-hearted steward as a pale, weary, grey-bearded man, sit sighing on the crumbling stones.

The Kolsassthal merges into the Weerthal and is hardly distinguished from it, and affords a sort of counterpart, though on more broken ground, to the Gnadenwald on the opposite side of the river. It is from this abundance of shady woods that its name is derived, through the old German *kuol*, cool, and *sazz*, a settlement. In the church, the altar-piece of the Assumption is by Zoller. The church of Wattens has an altar-piece by a more esteemed Tirolean artist, Schöpf; it represents S. Laurence, to whom the church is dedicated. The many forges busily at work making implements of agriculture, nails, &c., keep you well aware of the thrift and industry of the place; its prosperity is further supported by a paper manufactory, which has always remained in the hands of the family which started it in 1559, and supplies the greater part of Tirol. A self-taught villager, Joseph Schwaighofer, enjoyed some reputation here a few years ago as a guitar maker. The Wattenserthal, like the Kolsassthal, is also very woody, and contains some little settlements of charcoal-burners; but it is also diversified by a great many fertile glades, which are diligently sought out for pasture. At Walchen, where a few shepherds' huts are clustered at the confluence of two mountain streams, the valley is broken into two branches—one, Möls, running nearly due south into the Navisthal, by paths increasing in difficulty as you proceed; the other, Lizumthal, by the south-east to Hinterdux, passing at the Innerlahn the so-called 'Blue Lake,' of considerable depth.[1]

Following the road again, Volders is reached at about a mile from Wattens. As at the latter place, your ears are liberally greeted with the sounds of the smithy. Volders has quite a celebrity for its production of scythes; some ten or twelve thousand are said to be exported annually. The Post Inn affords tolerable quarters for a night or two while exploring the neighbourhood.

The prolific pencil of Schöpf has provided the church with an altar-piece of the Holy Family; though an ancient foundation, it does not present any object of special interest.

The Voldererthal runs beneath some peaks dear to Alpine climbers, the Grafmarterspitz, the Glunggeser, the Kreuzjoch, and the Pfunerjoch. Its entrance is commanded by the castles of Hanzenheim, sometimes called Starkelberg, from having belonged to a family of that name, and used as a hospital during the campaign

---

[1] Grimm (*Deutsche Sagen*, No. 492) gives an interesting legend of the Hasslacherbrunnlein (half way between Kolsass and Wattens) and of the resistance offered by the inhabitants of Tirol to the Roman invasion of their country.

of 1809; and Friedberg, which is still inhabited, having been carefully restored by the present owner, Count Albert von Cristalnigg. It was originally built in the ninth or tenth century, as a tower to guard the bridge; it gave its name to a powerful family, who are often mentioned in the history of Tirol. At the end of the thirteenth and beginning of the fourteenth centuries, it was one of the castles annexed by Friedrich *mit der leeren Tasche*. It contains also the Voldererbad, a mineral spring, which is much visited, but more conveniently reached by way of Windegg than through the valley itself.

In the Voldererwald is a group of houses, Aschbach by name, which belongs ecclesiastically to the parish of Mils, on the opposite side; and the following story is given to account for the anomaly:—At the time when the territory of Volders belonged parochially to Kolsass (it must have been before the year 1630, as it was that year formed into an independent parish), the neighbourhood was once ravaged by the plague. A farmer of Aschbach being stricken by it, sent to beg the spiritual assistance of the priest of Kolsass. The priest attended to the summons; but when he reached the threshold of the infected dwelling, and saw what a pitiable sight the sick man presented, his fears got the better of his resolution, and he could not prevail on himself to enter the room. Not to leave his penitent entirely without comfort, however, he exhorted him to repentance, heard his confession, and absolved him from where he stood; and then uncovering the sacred Host, bid him gaze on it in a spirit of faith, and assured him he should thereby receive all the benefit of actual Communion. The visit thus completed, he hurried back to Kolsass in all speed. Meantime the sick man, not satisfied with the office thus performed, sent for the priest of Mils, who, supported by apostolic charity, approached him without hesitation, and administered the sacred mysteries. Contrary to all expectation, the farmer recovered, resumed his usual labours, and in due course garnered his harvest. In due course also came round the season for paying his tithe. With commendable punctuality the farmer loaded his waggon with the sacred tribute, and started alacritously on the way to Kolsass. Any one who watched him might have observed a twinkle of his eye, which portended some unusual dénouement to the yearly journey. As he approached Kolsass the twinkle kindled more humorously, and the oxen felt the goad applied more vigorously. The pastor of Kolsass turned out to see the waggon approaching at the unusual pace, and was already counting the tempting sheaves of golden corn. To his surprise, however, his frolicsome parishioner wheeled round his team before he brought it to a stand, and then cried aloud, 'Gaze, Father! yes, gaze in faith on the goodly sight, and believe me, your faith shall stand you in stead of the actual fruition!' With that he drove his waggon at the same pace at which he had come, straight off to the pastor of Mils, at whose worthy feet he laid the tithe. And this act of 'poetical justice' was ratified by ecclesiastical authority as a censure on the pusillanimity of the priest of Kolsass, by the transfer of the tithing of Aschbach to the parish of Mils. I have met a counterpart of this story both in England and in Spain; so true is it, as Carlyle has prettily said, that though many traditions have but one root they grow, like the banyan, into a whole overarching labyrinth.

The stately *Serviten-kloster* outside Volders suggests another adaptation of this metaphor. From the root of one saint's maxims and example, what an 'over-arching labyrinth' of good works will grow up and spread over and adorn the face of the earth, even in the most distant parts. In the year 1590 there was born at Trent a boy named Hyppolitus Guarinoni, who was destined to graft upon Tirol the singular virtues of St. Charles Borromeo. Attached early to the household of the saintly Archbishop of Milan, Guarinoni grew up to embody in action his spirit of devotion and charity. By St. Charles's advice and assistance he followed the study of medicine, and took his degree in his twenty-fifth year. Shortly after, he was appointed physician in ordinary to the then ruler of Tirol, Archduke Ferdinand II. His fervent piety marked him as specially fit to be further entrusted with the sanitary care of the convent founded some years before by the Princesses Magdalen, Margaret, and Helena, Ferdinand's sisters, at Hall, and called the *Königliche Damenstift*.[1] All the time that

---

[1] The suppression of this and several other convents, in 1783, was a measure sufficiently unpopular to almost neutralize the popularity

was left free by these public engagements he spent by the bedsides of the poor of the neighbourhood. The care of the soul ever accompanied his care for their bodies, and many a wanderer owed his reconciliation with heaven to his timely exhortations. Just about this time the incursions of the new doctrines were making themselves felt in this part of Tirol, and some localities, which from their remoteness were out of the way of regular parochial ministrations, were beginning to listen to them. Guarinoni discovered this in the course of his charitable labours, for which no outlying *Sennerhütte* was inaccessible. In 1628 he obtained special leave, though a layman, from the Bishop of Brixen, to preach in localities which had no resident pastor; he further published a little work which he used to distribute among the people, designed to show them how many corporal infirmities are induced by neglect of the wholesome maxims of religion. Besides the restored unity of the faith in his country, two other monuments of his piety remain: the Church of St. Charles by the bridge of Volders, and the Sanctuary of Judenstein. In his moments of leisure it was his favourite occupation to commit to writing for the instruction of posterity the traditional details of the life of St. Nothburga, and of the holy child Andreas of Rinn, which were at his date

---

Joseph II. enjoyed as son of Maria Theresa. The suppression was not, however, accompanied by spoliation; the funds were devoted to provide a moderate stipend to a number of women of reduced circumstances belonging to noble families.

even more rife in the mouths of the peasantry of the neighbourhood than at present. He only died in 1654, having devoted himself to these good works for nearly half a century.

The church by which he endeavoured to bring under observation and imitation the distinguishing qualities of St. Charles, was erected on a spot famous in the Middle Ages as a bandit's den; the building occupied thirty-four years, and was consecrated but a short time before his death. Baron Karl von Fieger, from whom he bought the site, a few years later added to it the Servite monastery, which, though it exhibits all the vices of the architecture of its date, yet bears tokens that its imperfections are not due to any stint of means. Its three cupolas and other structural arrangements are designed in commemoration of the Holy Trinity—a mystery which is held in very special honour throughout Austria. In the decorations, later benefactors have carried on Guarinoni's intention, the acts of St. Charles being portrayed in the frescoes, completed in 1764, by which Knoller has earned some celebrity in the world of art for himself and for the church: they display his conversion from the stiffer German style of his master, Paul Trogger, to the Italian manner. That over the entrance conveys a tradition of St. Charles, predicting to Guarinoni, while his page, that he would one day erect a church in his honour; that of the larger cupola is an apotheosis of the saint. The picture of the high-altar sets forth the saint ministering to the plague-stricken; it is Knoller's boldest attempt at colouring.

Near the entrance door may be observed a considerable piece of rock built into the wall, entitled by the people '*Stein des Gehorsams*,'[1] its history being that at the time when the church was building it was detached from the rock above by a landslip, and threatened the workmen with destruction. Its course was arrested at the behest of a pious monk, who was overseeing the works.[2]

After passing the Servitenkloster a footpath may easily be found which leads to Judenstein and Rinn, the seat of one of the much-contested mediæval beliefs accusing the Jews of the sacrifice of Christian children. It may be better, in describing this stem of this banyan, to visit Rinn the further place first, and take Judenstein on our way back. The country traversed is well wooded, and further diversified by the bizarre outlines of the steeples of Hall seen across the river, while the mighty Glunggeser-Spitz rises 7,500 feet above you. It invites a visit for its amenity and its associations, though the relics of the infant Saint 'Anderle' are no longer there. His father died, it would seem, while he was a child in arms; his mother earned her living in the

---

[1] Stone of Obedience.

[2] I have met with another sprout of this banyan at the Monastery of the Sacro Speco in the Papal State, where a huge fragment of rock, so nicely balanced that it looks as if a breath might send it over the cliff, is pointed out as having stood still for centuries at the word of S. Benedict, who bid it '*non dannegiare i sudditi miei.*'

fields, and while she was absent used to leave her boy at Pentzenhof in charge of his godfather, Mayr. One day, when he was about three years old—it was the 12th July 1462—she was cutting corn, when suddenly she saw three drops of blood upon her hand without any apparent means of accounting for the token, one with which many superstitions were connected.[1] Her motherly instincts were alarmed, and, without an instant's consideration, she threw down her sickle and hurried home. A little field-chapel to St. Isidor the husbandman, St. Nothburga, and St. Andrew of Rinn, was subsequently built upon this spot. Arrived at Mayr's house, the forebodings of her anxious heart were redoubled at not finding her darling playing about as he was wont. The faithless godfather, taken by surprise at her unexpected return, only stammered broken excuses in answer to her reiterated inquiries. At last he exclaimed, thinking to calm her frenzy, 'If he is not here, here is something better—a hat full of golden pieces, which we will share between us.' He took down his hat, but to his consternation instead of finding it heavy with its golden contents, there was nothing in it but withered leaves! At this sight he was overcome with fear and horror; his speech forsook him, and his senses together, and he ended his days raving mad.

[1] Wolf, *Beiträge zur deutschen Mythologie*, vol. ii. pp. 17–21. Müller, *Niedersächsische Sagen*, p. 51. Müllenhoff, *Sagen der Herzogthümer Schleswig Holstein u Lanenburg*, p. 184.

The distracted mother, meantime, pursued her inquiries and perquisitions; but all she could learn was that certain Jews,[1] returning from their harvesting at Botzen, had over-tempted Mayr by their offers and persuaded him to sell the child to them, but with the assurance that he should come to no harm. Little reassured by the announcement, she ran madly into the neighbouring birchwood, whither she had learned they

---

[1] So strong is the prejudice in Tirol against Jews, that it is said to be most difficult to find any one who will consent to act the part of Judas in the Passion plays.

There is a very strong personal dislike to Judas throughout Tirol, and I have also heard that the custom of burning him in effigy occurs in various places. Karl Blind, in the article quoted above, (p. 3,) accounts for this custom in the following way: 'After the appearance of fermenting matter it was said' (in what he calls the germanic mythology) 'that there rose in course of time—even as in Greek mythology—first a half-human, half-divine race of giants, and then a race of Gods; the Gods had to wage war against the giants and finally vanquished them. Evidently the giants represent a torpid barren state of things in nature, whilst the Gods 'signify the sap and fulness of life which struggles into distinct and beautiful form. There was a custom among the Germanic tribes of celebrating this victory over the uncouth Titans by a festival, when a gigantic doll was carried round in Guy Fawkes manner and at last burnt. To this day there are traces of the heathen practice. In some parts of Europe, so-called Judas-fires, which have their origin in the burning of the doll which represented the giants or jötun. In some places, owing to another perversion of things and words, people run about on that fête-day shouting 'burn the old Jew!' The *jötun* was in fact, when Christianity came in, first converted into Judas and then into a Jew, a transition to which the similarity of the sound of the words easily lent itself.' No doubt *jötun* sounds very like *Juden* but not all coincidences are consequences, and it is quite possible that the old heathen custom had quite died out before that of burning Judas in effigy began, as it certainly had before Guy Fawkes began to be so treated. The same treatment of Judas' memory occurs, too, in Spain on the day before Good Friday.

had bent their steps, and there came upon the lifeless body of her treasure, hanging bloodless and mangled from a tree. A large stone near bore traces of having been used as a sacrificial stone, and the clothes, which had been rudely torn off, lay scattered about; the many wounds of his tender form showed by how cruel a martyrdom he had been called to share in the massacre of the Innocents.

His remains were tenderly gathered and laid to rest, and his memory held in affection by all the neighbourhood; nevertheless, though there were many signs of the supernatural connected with the event, it did not receive all the veneration it might have been expected to call forth.

About ten years later a similar event occurred at Trent, and the remains of the infant S. Simeon were treated with so great honour that the people of Rinn were awakened to an appreciation of the treasure they had suffered to lie in their churchyard almost unheeded.[1] The Emperor Maxmilian I. contemplated building a church over the spot where the martyrdom occured, hence call Judenstein. His intentions were frustrated by the knavery of the builder, and only a small chapel was built at this time; and though on occasion of its consecration the relics of the child martyr were carried thither in solemn procession, they were still for some time after preserved at Rinn. It was Hippolitus

---

[1] S. Simeon of Trent is commemorated in the Roman Breviary (on the 25th March). S. Andreas of Rinn has not received this honour.

Guarinoni to whom the honour is due of saving the spot from oblivion. The chisel of the Tirolese sculptor Nissl has set forth in grotesque design a group of Jews fulfilling their fearful deed. A portrait of Guarinoni was likewise hung up there. The relics were translated thither with due solemnity in 1678. An afflux of pilgrims was immediately attracted, and the numerous tablets which crowd the walls attest the estimation in which it has been held. Then the people began to remember the wonders that had surrounded it. The ghost of Godfather Mayr, which for two centuries had been frequently met howling through the woods, now seemed to have found its rest, for it was never more seen or heard. And they recalled how a beautiful white lily, with strange letters on its petals, had bloomed spontaneously on the holy infant's grave;[1] that when a wilful boy, Pögler by name, snapped the stem while they were still pondering what the unknown letters might mean, he had his arm withered; and further that for generations after, every Pögler had died an untimely or a violent death. How in like manner, for seven consecutive winters, the birch-tree, on which the innocent child's body was hung by his persecutors, put forth fresh green sprouts as if in

---

[1] Keller, in his *Volkslieder*, p. 242, gives an analogous legend of a poor idiot boy, who lived alone in the forest and was never heard to say any words but 'Ave Maria.' After his death a lily sprang up on his grave, on whose petals 'Ave Maria' might be distinctly read. It is a not unusal form of legend; Bagatta, *Admiranda orbis Christiani*, gives fifteen such.

spring, and how when a thoughtless woodman one day hewed it down for a common tree, it happened that he met with a terrible accident on his homeward way, whereof he died. It may well be imagined that where such legends prevailed Jews obtained little favour; so that to the present day it is said there is but few Jew families settled among them, though they are numerous and influential in other parts of the Austrian dominion.[1]

[1] The ballad concerning the analogous English Legend of Hugh of Lincoln seems to demand to be remembered here:—

HUGH OF LINCOLN

(SHOWING THE CRUELTY OF A JEW'S DAUGHTER).

A' the boys of merry Lincoln,
  Were playing at the ba',
And up it stands him, sweet Sir Hugh,
  The flower among them a'.

He kicked the ba' there wi' his feet,
  And keppit it wi' his knee,
Till even in at the Jew's window,
  He gart the bonny ba' flee.

'Cast out the ba' to me, fair maid,
  Cast out the ba' to me;'
'Never a bit,' says the Jew's daughter,
  'Till ye come up to me.'

'Come up, sweet Hugh! come up, dear Hugh!
  Come up and get the ba';'
'I winna come, I minna come,
  Without my bonny boys a'.'

She's ta'en her to the Jew's garden,
  Where the grass grew long and green;
She's pu'd an apple red and white,
  To wyle the bonny boy in.

Another memory yet of Hippolitus Guarinoni lingers in the neighbourhood. By a path which branches off near Judenstein to the left (going *from* Volders and following the stream), the Volderbad is reached; a sulphur spring discovered and brought into notice by him, and now much frequented in summer,

> When bells were rung and mass was sung,
>   And every bairn went home;
> Then ilka lady had her young son,
>   But Lady Helen had none.
>
> She row'd her mantle her about,
>   And sair, sair, 'gan to weep;
> And she ran into the Jew's house
>   When they were all asleep.
>
> . . . . .
>
> 'The lead is wondrous heavy, mither,
>   The well is wondrous deep;
> A keen penknife sticks in my heart,
>   Tis hard for me to speak.'
>
> 'Gae hame, gae hame, my mither dear,
>   Fetch me my winding-sheet;
> And at the back of merry Lincoln,
>   'Tis there we twa shall meet.'
>
> Now Lady Helen she's gane hame,
>   Made him a winding-sheet;
> And at the back o' merry Lincoln,
>   The dead corpse did her meet.
>
> And a' the bells o' merry Lincoln
>   Without men's hands were rung;
> And a' the books o' merry Lincoln,
>   Were read without men's tongue;
> Never was such a burial
>   Since Adam's days begun.

perhaps as much for its pleasant mountain breezes as for the medicinal properties of the waters.

There is another interesting excursion which should be followed before reaching Innsbruck, but it is more easily made from Hall than from Volders, though still on the right bank of the Inn. The first village on it is Ampass, a walk of about four miles from Hall through the most charming scenery; it is so called simply as being situated on a pass between the hills traversed on the road to Hall. Then you pass the remains of the former seat of the house of Brandhausen; and following the road cut by Maria Theresa through the Wippthal to facilitate the commerce in wine and salt between Matrei and Hall, you pass Altrans and Lans, having always the green heights of the Patscherkofl smiling before you, an easy ascent for those who desire to practise climbing, from Lans, where the *Wilder Man* affords possible quarters for a night.[1] A path branching off from the Mattrei road leads hence to Sistrans, a village whose church boasts of having been embellished by Claudia de' Medici. Its situation is delightful; the green plain is strewed with fifteen towns and villages, including Hall and Innsbruck, and behind these rise the great range of alps, while on the immediate foreground is the tiny Lanserse which will afford excellent *Forellen* for luncheon. The bed of this same Lansersee, it is said, was once covered

---

[1] There is a carriage-road reaching nearly to the top of the Lanserkopf.

with a flourishing though not extensive forest, its wood the only substance of a humble peasant, who had received it from his fathers. A nobleman living near took a fancy to the bit of forest ground, but instead of offering to purchase it, he endeavoured to set up some obsolete claim in a court of law. The judge, afraid to offend the powerful lord, decided in his favour. The poor man heard the sentence with as much grief at the dishonour done to his forefathers' honour as distress at his own ruin. 'There is no help for me on earth, I know,' said the poor man. 'I have no money to make an appeal. I may not contend in arms with one of noble blood. But surely He who sitteth in heaven, and who avenged Naboth, will not suffer this injustice. As for me, my needs are few; I refuse not to work; the sweat of my brow will bring me bread enough; but the inheritance of my fathers which I have preserved faithfully as I received it from them, shall it pass to another?' and in the bitterness of his soul he wept and fell asleep; but as he slept in peace a mighty roaring sound disturbed the slumbers of the unjust noble; it seemed to him in his dream as though the foundations of his castle were shattered and the floods passing over them. When they awoke in the morning the forest was no more to be seen—a clear calm lake mirrored the justice of heaven, and registered its decree that the trees of the poor man should never enrich the store of his unscrupulous neighbour.

Sistrans was once famous for a champion wrestler who had long carried off the palm from all the country round; but like him of Schwatz, he was not content with his great natural strength; he was always afraid a stronger than he might arise and conquer him in turn; and so he determined to put himself beyond the reach of another's challenge. To effect this he arranged with great seeming devotion to serve the Mass on Christmas night; and while the priest's eye was averted, laid a second wafer upon the one that he had had laid ready. The priest, suspecting nothing, consecrated as usual; and then at the moment of the *Wandlung,* when the priest was absorbed in the solemnity of his act, as he approached to lift the chasuble he stealthily abstracted the Host he had surreptitiously laid on the altar. The precious talisman carefully concealed, he bound it on his arm the instant Mass was over; and from that day forth no one could stand against him. And not only this, but he had power too in a multitude of other ways. Had anyone committed a theft, it needed but to consult our wrestler; if he began saying certain words and walking solemnly along, immediately, step by step, were he far or near, the thief, wherever he was, was bound by secret and resistless impulse to tread as he trod, and bring back the booty to the place whence he had taken it. Was anyone's cattle stricken with sickness, it needed but to call our wrestler; a few words solemnly pronounced, and the touch of his potent arm, sufficed to restore the beast to perfect health. Moreover, no bird could escape his snare, no fox or hare or chamois outrun him for swiftness.

Thus all went well; he had played a bold stake, and had won his game. But at last the time came for him to die. Weary of his struggles, and even of his successes, our wrestler would fain have laid his head to rest under the soft green turf of the field of peace, by the wayside of those who pass in to pray, and lulled by the sound of the holy bells. But in vain he lay in his bed; death came not. True, there were all his symptoms in due force—the glazed eye and palsied tongue and wringing agony; but for all that he could not die. At last, the priest, astonished at what he saw, asked him if he had not on his conscience some sin weighty above the wont, and so moved him to a sense of penance that he confessed his impiety with tears of contrition; and it was not till he had told all, and the priest had received the sacred particle he had misused, that, shriven and blessed, his soul could depart in peace. There is a spot outside Sistrans called the *Todsünden-marterle*, but whether it has any connection with this tradition, or whether it has one of its own, I have not been able to learn.

A couple of hours further is the pilgrimage chapel of Heiligenwasser, which is much visited both by the pious and the valetudinarian. Its history is that in

1606 two shepherd boys keeping their father's herd upon the mountains lost two young kine. In vain they sought them through the toilful path and beneath the burning sun; the kine were nowhere to be found. At last in despair of any further labour proving successful, they fell on their knees and prayed with tears for help from above. Then a bright light fell upon them, and the *Gnadenmutter* appeared beside them, and bid them be of good cheer, for the cattle were gone home to their stall; moreover she added, 'Drink, children, for the day is hot, and ye are weary with wandering.' 'Drink!' exclaimed the famished children, 'where shall we find water? there is no water near!' but even as they spoke the *Gnadenmutter* was taken from their sight, but in the place where the light surrounding her had shone there welled up a clear and bubbling stream between the rocks, which has never ceased to flow since. The boys went home, but had not the courage to tell how great a favour had been bestowed on them; yet they never went by that way without turning to give glory to God, and say a prayer beneath the holy spring.

Fifty years passed. One of them was an infirm old man, and no longer went abroad so far, the other was attended in his labours by the son of a neighbour, a lad who had been dumb from his birth. When the lad saw the herdsman kneel down by the spring and drink and pray, he knelt and drank and prayed too; when lo! no sooner had the water passed his lips than he found he had the power of speech like any other. The narration of the one wonder led to that of the other. The people readily believed, and before the year was out a chapel had been raised upon the spot.

CHAPTER VIII.

NORTH TIROL—THE INNTHAL.
INNSBRUCK.

Many centuries have been numbered,
Since in death the monarch slumbered
By the convent's sculptured portal,
  Mingling with the common dust ;
But his good deeds, through the ages
Living in historic pages,
Brighter grow and gleam immortal,
  Unconsumed by moth or rust.
LONGFELLOW.

I SHALL not easily forget my first greeting at Innsbruck. We had come many days' journey from the north to a rendezvous with friends who had travelled many days' journey from the south; they were to arrive a week earlier than we, and were accordingly to meet us at the station and do the hosts' part. But it happened that the station was being rebuilt, and the order of 'No admittance except on business' was strictly enforced. The post-office was closed, being 'after hours,' and though the man left in charge, with true Tirolean urbanity, suffered us to come in and turn over the letters for ourselves, we failed to find the one conveying the directions we sought. So with no fixed advices to guide us, we wandered through the mountain capital in search of a chance meeting. We had nearly given up this attempt in its turn in despair of success, when 'Albina,' a little white Roman *lupetto* dog, belonging to the friends of whom we were in search, came bounding upon me. It was more than two years since I had taken leave of her in the Eternal City, but her affectionate sympathy was stronger than time or distance; and here, far from all aid in the associations of home, and while the rest of her party were yet a great way off—almost out of sight—she had spied me out, and came to give her true and hearty greeting.

It is a pleasant association *with* Innsbruck, a revelation of that pure and lasting love which dog-nature seems to have been specially created to convey; but it was not *of* Innsbruck. Innsbruck—*Schpruck*, as the indigenous call it—though the chief, is the least Tirolean town of Tirol. It apes the airs and vices of a capital, without having the magnificence and convenience by which they are engendered.

There is a page of Tirol's history blotted by a deed which Innsbruck alone, of all Tirol, could have committed, and which it indeed requires its long and otherwise uniformly high character for both exceeding hospitality and exceeding loyalty to cancel. The subject of it was its own Kaiser Max, whose prudence in governmental details and gallantry in the field and in the chase had raised him in the popular mind to the position of a hero. When he had come to them before, in his youth, in his might, and in his imperial pomp, he had been sung and fêted. The people had acclaimed him with joy, and his deeds were a very household epic; while he in turn had extended their borders by conquest, and their privileges by concessions. But now he had come back to them, worn out with war and cares and age. He felt that his end was near, and it was to Tirol, with which he had always stood in bonds of so much love, that he turned to spend his few declining years. But Innsbruck, when it saw him thus, seems to have forgotten his prowess and his benefits, and to have remembered only a pitiful squabble about payment of the score for the maintenance of his household at his last visit. A ruler who had spent himself in bettering the condition of his people might well, in the days of his weariness and sadness of heart, have expected to meet with more liberality at their hands; but from Innsbruck, where—little obscure provincial town as it was—he had so often held his court, which had been raised in importance and singularly enriched by royal marriages and receptions and other costly ceremonies celebrated there at his desire, and which by his example and instigation had become the residence of many nobles who had learnt under his administration to value peaceful study above the pursuit of war—from Innsbruck he had most of all to expect. And yet on this occasion, as he lay ailing and restless on his couch, the neighing and tramping of his horses disturbed his fitful slumbers; and rising in the early dawn to ascertain the cause, he beheld the team which had brought him from the Diet at Augsburg, left out unfed and untended in the streets, because the people said he should not run up another score with them. With a moderation he would not perhaps have practised in his younger days, he quietly went on his way, to die at Wels on the Traun.

I have often pictured the pale sad face of the old Emperor as he turned from that sight, and thought of the sickness of his heart as one of history's most touching lessons of the world's inconstancy. Perhaps it predisposed me against Innsbruck; perhaps I was inclined to be a little unjust; but, at all events, it prepared me not to be surprised if its people should prove more sophisticated than their fellow-countrymen. It was quite what I expected, therefore, when I was told that in the older inns of the class wherein one generally finds a refreshing hospitality and primitiveness, the absence of comfort was not compensated by corresponding simplicity of manners.

In the Oesterreichisher Hof, one of those provincial pieces of pretentiousness which those who travel to learn the characteristics of a country should, under ordinary circumstances, avoid, we found the pleasantness of its situation sufficient to make us forget all else; and indeed, considered as a copy of a Vienna hotel, it is not a bad attempt. There is a room which on Sundays is set apart for an English service. On a subsequent visit we found a large new hotel (Europa), rather near the railway station, preferable to it in some respects, and there are many others besides.

I have spoken of the pleasant situation, and our apartment was situated so as fully to enjoy it; we had to ourselves a whole suite of little rooms, with a separate corridor running along the back of them, from the windows of which we could make acquaintance, under the alternating play of sunshine, moonbeam, or lightning, with the range of mountains which wall in Tirol. The Martinswand and Frauhütt, with their romantic memories; the Seegruben-spitze and the Kreuz-spitze, rugged and wild; the grand masses of the Brandjoch and the lesser Solstein, and the greater Solstein already wearing a lace-like veil of snow; while the quaint copper cupolaed towers of Innsbruck conceal the Rumerjoch and the Kaisersaüle; and in the front of the picture, the roofs with their wooden tiles afford a view of the mysteries of apple-drying, and a thousand other local arts of domestic economy. If our furniture was not of the most elegant or abundant, it was all the more in keeping with such wild surroundings.

The character of the town itself partakes of the same mixture of quaint picturesqueness with modern pretension which I have already observed in that of the people and the hotel. The *Neustadt*, as the chief street is called, remarkable for its width, tidiness, and good paving, is no less so for its old arcades in one part, and the steep gables in another, and the monuments of faith which adorn its centre line. At one end it is closed in by the stern gaunt mountain, at the other by Maria Theresa's triumphal arch. There are other streets again, straight, modern, and uniform; the Museum Strasse, and the Karl Strasse, and the Landhaus Gasse,[1] but you soon come to an end of them; and then you find yourself in a suburb of most primitive quality; your progress arrested, now by the advance of the iron road, now by the placid gentle Sill, now by the proudly flowing Inn. The mediæval history of Innsbruck is signalized by a number of fires which destroyed many of its antiquities. To the first of these it owes the suggestion that the town needed a water supply, acted upon by Meinhard II., and the monks of Wilten, in the formation of the *Kleine Sill*, which continues still as useful as ever; but other fires again and again laid it in ashes, so that very little of really old work survives, though there are many foundations of early date, the buildings of which have been again and again rebuilt. The very oldest of these is the monastery of Wilten, now a suburb a little way outside the *Triumphpforte*, originally the seat of the suzerains who created the town.

The history of its origin is one of the most remarkable myths of the country, and is a very epitome of the history of the conflict of Heathendom with Christendom.

[1] The best shops are in the Franziskanergruben.

The Romans had found here a flourishing town even in their time, and they made of it an important station, calling it Valdidena, whence its present name; coins and other relics of their sojourn are continually dug out of the soil. Tradition has it, however, that Etzel (Attila) laid the city in ruins on his way back from the terrible battle of Chalons. It continued, nevertheless, to be a convenient and consequently frequented station of the intercourse between the banks of the Po and the Rhine. When Dietrich von Bern (Theodoric of Verona) announced his expedition against Chriemhilde's Garden of Roses at Worms, one of the mightiest who responded to his appeal, and who did him the most signal service in taking the Rose-garden, was Heime, popularly called Haymon, a giant 'taller and more powerful than Goliath.' Returning in Theodoric's victorious train, he came through Tirol. As he approached Valdidena he found his passage barred by another giant named Thyrsus, living near Zirl, who has left his name to the little neighbouring hamlet of Tirschenbach. Thyrsus had heard of Haymon's prowess, and as his own had been unchallenged hitherto, he determined to provoke him to combat. Haymon was no less fierce than himself, and scarcely waited for his challenge to rush to the attack. But anyone who had looked on would have guessed from the first moment on which side the advantage would fall. Thyrsus was indeed terrible of aspect; higher in stature than Haymon, his shaggy hair covered a determined brow; his hardy skin was bronzed by exposure to weather and lying on the rocks; his sinews were developed by constant use, and their power attested by the tree torn up by the roots which he bore in his hand for a club; at each footfall the ground shook, for he planted his feet with a sound of thunder, and his stride was from hill to hill. But Haymon's every movement displayed him practised in each art of attack and defence. Less fierce of expression than Thyrsus, his eyes were ever on the watch to follow every moment of his antagonist, and like a wall of adamant he stood receiving all his thrusts with a studied patience, giving back none till his attacker's strength was well-nigh exhausted. *Then* he fell upon him and slew him. An effigy of the two giants yet adorns the wall of the wayside chapel at Tirschendorf.

Haymon was still in the prime of manhood, being about thirty-five, and this was but one of his many successful combats. Nevertheless, it was destined to be his last, for a Benedictine monk of Tegernsee coming by while he was yet in the first flush of victory, succeeded so well in reasoning with him on the worthlessness of all on which he had hitherto set his heart, and on the superior attractions of a higher life, that he then and there determined to give up his sanguinary career, and henceforth devote his strength to the service of Christ.

In pursuance of this design he determined to build with his own hands a church and monastery on the site of the ruined town of Valdidena, by the banks of the

Sill. With his own hands he quarried the stone and felled the timber; but in the meantime the Evil One in the form of a huge dragon had taken possession of the place. Never did he let himself be seen; but when he came to lay the foundation, Haymon found every morning that whatever work he had done by day, the dragon had destroyed by night. Then he saw that he must watch by night as well as work by day, and by this means he discovered with what manner of adversary he had to deal. The dragon lashed the ground with his tail in fury, just as the wild wind stirs up the sea, and filled the air with the smoke and sparks he breathed out of his mouth. Haymon saw that with all his strength and science he could not overcome so terrible an enemy; nevertheless, he did not lose heart, but commended himself to God. Meantime, the streaks of morn began to appear over the sky, and at sight of them the dragon turned and fled. Haymon perceived his advantage, and pursued him; by-and-by the rocks bounding the path contracted, and at last they came to the narrow opening of a cave. As soon as the dragon had got his head in and could not turn, Haymon raised his sword with a powerful swing, and calling on God to aid his stroke, with one blow severed the monster's head from the trunk. As a trophy of his feat, he cut out the creature's sting, which was full two feet long, and subsequently hung it up in the Sanctuary, and something to represent it is still shown in the church of Wilten.

After this, the building went on apace; and when it was completed, he took up a huge stone which had been left over from the foundation of the building, and flung it with the whole power of his arm. It sped over the plain for the space of nearly two miles, till it struck against the hill of Ambras, and rolled thence down again upon the plain, 'where it may yet be seen;' and with all the land between he endowed the monastery. Then he called thither a colony of Benedictines to inhabit it, and himself lived a life of penance as the lowest among them for eighteen years; and here he died in the year 878. Another benefit which he conferred on the neighbourhood was rebuilding the bridge of Innsbruck.[1] Tradition says he was buried on the right hand side of the high altar, and even preserves the following rough lines as his epitaph:—

> Als Tag und Jahr verloffen war
> Achthundert schon verstrichen
> Zu siebzig acht hats auch schon g'macht
> Da Heymons Tod verblichen.
> Der tapfere Held hat sich erwählt
> Ein Kloster aufzuführen
> Gab alles hinein, gieng selbst auch drein,
> Wollte doch nicht selbst regieren.
> Hat löblich gelebt, nach Tugent gstrebt
> Ein Spiegel war er allen;
> Riss hin riss her, ist nicht mehr er,
> Ins Grab ist er hier g' fallen.

Many fruitless searches have been made for his body; the last, in the year 1644, undermined great

---

[1] Grimm, *Deutsche Sagen*, No. 139.

part of the wall of the church, and caused its fall. The popular belief in the existence of the giants Haymon and Thyrsis has found a forcible expression nevertheless in two huge wooden figures, placed at the entrance of the Minster Church.

The parish church of Wilten has a more ancient and curious relic in the *Mutter Gottes unter den vier Säulen*,[1] of which it is said, that the Thundering Legion having been stationed at Valdidena about the year 137, had this image with them; that on one occasion of being ordered on a distant expedition they buried it under four trees, and never had the opportunity of recovering it. That when Rathold von Aiblingen made his pilgrimage to Rome, he brought back with him the secret of its place of concealment, exhumed it, set it up on the altar under a baldachino with four pillars, where it has never ceased to be an object of special veneration. This received a notable encouragement when Friedrich *mit der leeren Tasche*, wandering in secret through the country with his trusty Hans von Müllinen after the ban of the empire had been pronounced against him, knelt before this shrine, and prayed a blessing on his unchanging devotion to it. The sequel made him believe that his prayer was heard; and when he was once more established in his possessions, he caused himself and his friend to be portrayed kneeling at the shrine to seek protec-

---

[1] Under four pillars.

tion under the fostering mantle of the Virgin, and had the picture hung on the wall of the church opposite.

The name of Innsbruck first occurs in a record of the year 1027, on occasion of a concession granted to the chapel of *S. Jakob in der Au*—S. James's in the Field—probably the spot on which the stately Pfarrkirche now stands. Prior to this, the little settlement of inhabitants, whom the commerce between Germany and Italy had gathered round the Inn-bridge, could only satisfy the obligation of the Sunday and Holy-day mass by attendance in the church of Wilten; now, the faculty was granted to their own little chapel.

Its situation made it a convenient entrepôt for many articles of heavy merchandise, and, as years went by, a dwelling-place of various merchants also. All this time it was a dependency of the monks of Wilten. In 1180, Berthold II. von Andechs, acquired from them by treaty certain rights over the prospering town. His successor, Otho I., surrounded it with walls and fortifications, and built himself a residence, on the entrance of which was chiselled the date of 1234, and the inscription,—

>   Dies Haus steht in Gottes Hand
>   Ottoburg ist es genannt.

And on the same spot, in an old house overlooking the river Inn, some remains of this foundation may be traced, to which the name of Ottoburg still attaches.

In 1239 it was treated to the privilege of being

the only dépôt for goods between the Ziller and the Melach; other concessions followed, maintaining its ever-rising importance. In 1279 Bruno, Bishop of Brixen, consecrated a second church, the Morizkapelle, in the Ottoburg. But though both its temporal and spiritual lords appear to have encouraged its growth by every means in their power, and though there are records of occasional noble gatherings within its precincts, it was not till after the cession of Tirol to Austria by Margaretha Maultasch that the convenience of its central situation, and its water communication by the Inn and Danube with other towns of the empire, suggested its adoption as the seat of government of the country.

The fidelity of the towns-people to Duke Rudolf IV. of Austria at the time of a Bavarian invasion, elicited a further outpouring of privileges from their ruler, putting beyond all dispute in a short time the priority of Innsbruck over all the towns of Tirol.

Friederich *mit der leeren Tasche* made it his residence, and his base of operations for reducing the Rottenburgers and other powerful nobles, who during the late unsettled condition of the government had set at naught his power and oppressed the people. In this he received the warmest support of the Innsbruckers, which he in turn repaid by granting all their wishes.

The singular loyalty of the Tirolese, and their good fortune in having been generally blessed with upright and noble-minded rulers, make their annals read like a continuous heroic romance. The deeds of their princes have for centuries been household words in every mountain home of Tirol. None have had a deeper place in their hearts than the fortunes of *Friedl*, and never was any man more fortunate in his misfortunes. Before they yet knew what manner of prince he was, the ban of the empire had made him a penniless wanderer. Reduced to a condition lower than their own, the peasants wherever he passed gathered round him, and swore to stand by him, and concealed his hiding-places with the closest fidelity. One night he came weary and wayworn to Bludenz in Vorarlberg, seeking shelter before the impending storm. The night-watch had the closest orders to beware of strangers, for an incursion of the imperial army was expected, and every stranger might be a spy; no entreaty of Friedl on his friend Hans could shake his obedience to orders. When the Prince declared who he was, the man said, 'Would it were Friedl indeed!' but added that he would not be taken in by the pretence, however well devised. At last the outcast obtained from him that he would send for an innkeeper to whom he was known. Mine host at once recognised his sovereign, and received him with joy. The *Thorwachter* trembled when he found what he had done, but Frederick commended his steadfastness heartily, and invited him to dine at his table next day. While he was here, the Emperor summoned the burghers to give up his prisoner; but the Bludenzers sent answer that 'they had sworn fealty to Duke Frederick and the House of Austria, and they would not break their oath.' This spirited reply would probably have brought an army to their gates had Frederick remained among them; but in order to save them from an attack, for which they were little prepared, he took his departure,—by stealth, or they would not have suffered him to depart, even for their own safety's sake. At other times he would earn his day's food by manual labour before he disclosed to his entertainers who he was, and then he would only partake of the same frugal fare, and the same hard lodging, as the peasants who received him. By these means he became deeply endeared to the people, who thus knew he was one who felt for their privations, and shared their feelings and opinions, and did not treat them with supercilious contempt like one of the nobles.

When by these wanderings Frederick had discovered how deeply the people loved him, he arranged with the owner of the Rofnerhof in the Oetztal a plan by which, on occasion of a great fair at Landeck, always crowded by people from all the country round, he appeared in the character of principal actor in a peasant-comedy, which set forth the sufferings of a prince driven from his throne by cruel enemies, wandering homeless among his people, then calling them to arms, and leading them to victory. The excitement of the people at the representation exceeded his highest expectations. Loud sobs and cries accompanied his description of the Prince's woes; but when he came to sing of the people following their prince's call to arms, their ardour became quite irresistible. The enthusiasm was contagious; Frederick could no longer contain himself; he threw off his disguise, and declared himself their Friedl. It needed no more; unbidden they proffered their allegiance and their vows to defend his rights to the last drop of their blood. The enthusiasm of the Landeckers soon spread over the whole country; and when the Emperor Sigismund and Ernst der Eiserne and Frederick's other foes found his people were as firm as their own mountains in his defence, they gave up the attempt at further persecution, and concluded a truce with him.

In his prosperity he did not forget the peasants who had stood by him so loyally. While he tamed the power of their oppressors, he did all he could to lighten their burdens; and to many, who had rendered him special service, he marked his gratitude by special favours. Thus, to Ruzo of the Rofnerhof he granted among other privileges the right of asylum on his demesne, which was put in use down to the year 1783. We have already seen his conflict with Henry of Rottenburg,[1] and in the same way he tamed the overgrown power of other nobles. In the course of our wanderings we shall often find the popular hero's name stored up in the people's lore.

In connection with Innsbruck, he is well known to the most superficial tourist as the builder of the *Goldene Dachl-gebäude*.

[1] See p. 69.

And what is the *goldene Dachl-Gebäude*?—It is a most picturesque addition to, and almost all remaining of, what in his time was the *Fürstenburg*, or princely palace, having a roof of shining gilt copper tiles, sufficiently low to be in sight of the passer-by; but the account the best English guide-book gives the tourist of its origin is so wanting in the true appreciation of Friedl's character, that I am fain to supply the Tirolese version of it. The above account says that it was built in 1425 ' by Frederick, called in ridicule "Empty Purse," who, in order to show how ill-founded was the nickname, spent thirty thousand ducats on this piece of extravagance, which probably rendered the nickname more appropriate than before.' Now, to say that he *was called* ' Empty Purse' thus vaguely would imply that it was a name given by common consent, and generally adopted. To say that he built the Golden Roof only to show that such a nickname was ill-founded, is simply to accuse him of arrogance. To treat it as an extravagance which justified the accusation, is to convict him of folly.

But the government of Frederick [1]—which is felt even yet in the present independent spirit of Tirol, which consolidated the country and made it respected, which set up the dignity of the *Freihof* and the *Schildhof* the foundation of a middle class as a dam against the encroachments of the nobility on the peasantry,

---

[1] Of the earlier history of Tirol we shall have to speak when we come to Schloss Tirol and Greifenstein.

which yet lives on in the hearts of the people, was an eminently prudent administration, and the story does not fit it. If, instead of resting satisfied with this compendious but flippant account, you ask the first true Tirolese you meet to expound it, he will tell you that Friedl had grown so familiar with peasant life that he despoiled himself to better the condition of his poorer subjects, not only by direct means, but by his expeditions in their defence, and also in forbearing to exact burdensome taxes. The nickname was not given him by general consent; nor at all, by the people; it was the cowardly revenge of those selfish nobles who could not appreciate the abnegation of his character. Frederick saw in it a reproach, offered not so much to himself as to his people; it seemed to say that the people who loved him so well withheld the subsidies which should make him as grand as other monarchs. To disprove the calumny, and to show that his people enabled him to command riches too, he made this elegant little piece of display, which served also to adorn his good town of Innsbruck; but he did not on that account alter his frugal management of his finances; so that when he came to die, though he had made none cry out that he had laid burdens on them, he yet left a replenished treasury.[1]

This is still one of the notable ornaments of Innsbruck. The house is let to private families, but

[1] Consult Zoller, *Geschichte der Stadt Innsbruck*; and Staffler, *das Deutsche Tirol*.

the 'gold-roofed' *Erker*, or oriel, is kept up as a beloved relic almost in its original condition. There is a curious old fresco within, the subject of which is disputed; and on the second floor there is a sculptured bas-relief, representing Maximilian and his two wives, Mary of Burgundy and Maria Bianca of Milan, and the seven coats of arms of the seven provinces under Maximilian's government.

Sigismund 'the Monied,' Frederick's son and successor (1430-93), is more chargeable with extravagance,[1] but his extravagance was all for the advantage of Innsbruck. The reception he gave to Christian I., King of Denmark, when on his way to Rome, is a striking illustration of the resources of the country in his time. Sigismund went out to meet him at some miles' distance from the capital, with a train of three hundred horses, all richly caparisoned; his consort (Eleanor of Scotland) followed with her suite in two gilt carriages, and surrounded by fifty ladies and maidens on their palfreys. The King of Denmark stayed three days; every day was a festival, and the magnificent dresses of the court were worthy of being specially chronicled. There seems to have been no lack of satin and velvet and ermine, embroidery, and fringes of gold-work.

Nor was mental culture neglected; for we find mention, at the same date, of public schools governed

---

[1] See p. 146.

by 'a rector,' which would seem to imply that they had something beyond an elementary character. The impulse given to commerce by the working of the silver-mines also had the effect of causing some of the chief roads of the country to be made and improved. The most lasting traces of Sigismund's reign, however, are the ruined towers which adorn the mountain landscapes. Wherever we go in Tirol, we come upon some memory of his expensive fancy for building isolated castles as a *pied à terre* for his hunting and fishing excursions, still distinguished by such names as Sigmundskron, Sigmundsfried, Sigmundslust, Sigmundsburg, Sigmundsegg, and which we shall have occasion to notice as we go along. His wars were of no great benefit to the country, but his command of money enabled him to include Voralberg within his frontier. Sigismund was, however, entirely wanting in administrative qualities. This deficiency helped out his extravagance in dissipating the whole benefit which might have resulted to the public exchequer from the silver-works of his reign; and at last he yielded to the wholesome counsel of abdicating in favour of his cousin Maximilian.

Maximilian (1493-1519) is another of the household heroes of Tirol. Even after he was raised to the throne of empire he still loved his Tirolean home, and his residence there further increased the importance of the town of Inusbruck. He built the new palace in the Rennplatz, called the Burg, which was completed for his marriage with Maria Bianca, daughter of Galeazzo Maria Sforza, of Milan. Splendid was the assemblage gathered in Innspruck for this ceremonial. Three years later it was further astonished by the magnificence of the Turkish Embassy; and the discussion of various treaties of peace were also frequently the means of adding brilliancy to the court, and prosperity to the town. His other benefits to the city, and Innsbruck's unworthy return to him, I have already mentioned in the beginning of this chapter.

Many a fantastic *Sage* is told of Maximilian in the neighbourhood, which we shall find in their due places. The fine hunting-ground Tirol affords was one of its greatest attractions for him; it led him, however, to introduce certain game-laws, and this was one principal element in bringing about the decline of his popularity in the last years of his life. At his death this disaffection broke out, and caused one of the most serious insurrectionary movements which have disturbed the even tenour of Tirolese loyalty. To this was added the influence of Lutheran teaching, the effects of which we have seen in the Zillerthal.

This spirit of discontent had time to gain ground during the first years of Maximilian's grandson and successor, Charles Quint, whose immensely extended duties drew his attention off from Tirol. Very shortly after his accession, however, he made over the German hereditary dominions, including Tirol, to his brother Ferdinand, who established his family in this country.

His wise administration and prudent concessions soon conciliated the people; though severe measures were also needed, and the year 1529 was signalized in Innsbruck by some terrible executions. These were forgotten when, in the year 1531, Charles Quint, returning victorious from Pavia, on his way to Augsburg stayed and held court at Innsbruck; Ferdinand met him on the Brenner pass, and accompanied him to the capital. When Charles reached the Burg, Ferdinand's children received him at the entrance; and the tenderness with which he greeted and kissed them was remarked by the people, on whom this token of homely affection had a powerful effect. Electors and princes, spiritual and temporal, came to pay their homage to the Emperor; and Innsbruck was so filled with the titled throng, that the Landtag had to remove its session to Hall. Ferdinand's other dominions, and the question of the threatened war with Turkey, necessitated frequent absences from Innsbruck. During one of these (in 1534) the Burg was burnt down, and his children were only rescued from their beds with difficulty. The great Hall, called the *goldene Saal*, and the state bedroom, which was so beautifully ornamented that it bore the title of *das Paradies*, were all reduced to ashes. In 1541 Innsbruck was once more honoured by a visit of the magnificent Emperor; and again, ten years later, he took up his residence there, that he might be near the Session of the Council of Trent. It was while he was living here peacefully in all confidence, and almost unattended, that Maurice, Elector of Saxony, having suddenly joined the Smalkald League, treacherously attempted to surprise him, marching with a considerable armed force through pass Fernstein. Charles, who was laid up with illness at the time, was enabled by the loyal devotion of the Tirolese to escape in the night-time and in a storm of wind and rain, being borne in a litter over the Brenner, and by difficult mountain paths through Bruneck into ' Carinthia. Maurice, baffled in his scheme, exercised his vengeance in plundering the imperial possessions, while his followers devastated the peasants' homes, the monastery of Stams, and other religious houses that lay in their way. The sufferings of the Tirolese on this occasion doubtless tended to confirm them in their aversion for the Lutheran League. Maurice's end was characteristic, and the Tirolese, ever on the look-out for the supernatural, were not slow to see in it a worthy retribution for his treatment of their Emperor. Albert of Brandenburg refused to join in the famous Treaty of Passau, subsequently concluded by Maurice and the other Lutheran leaders with the Emperor. This and other differences led to a sanguinary struggle between them, in the course of which Maurice was killed in battle at Sieverhausen.

Ferdinand the First's reign has many mementos in Innsbruck. He built the Franciscan church, otherwise called the *heiligen Kreuzkirche* and the *Hofkirche*, which, tradition says, had been projected by his grandfather, Kaiser Max, though there is no written record of the fact; and he raised within it a most grandiose and singular monument to him, which has alone sufficed to attract many travellers to Tirol. The original object of the foundation of the church seems to have been the establishment of a college of canons in this centre, to oppose the advance of Lutheran teaching. It was begun in 1543, the first design having been rejected by Ferdinand as not grand enough, and consecrated in 1563. He seems to have been at some pains to find a colony of religious willing to undertake, and competent to fulfil, his requirements; and not coming to an agreement with any in Germany or the Netherlands, ultimately called in a settlement of Franciscans from Trent and the Venetian provinces, consisting of twenty priests and thirteen lay-brothers. The chief ornaments of the building itself are the ten large—but too slender —red marble columns, which support the *plateresque* roof. The greater part of the nave is taken up with Maximilian's monument—cenotaph rather, for he lies buried at Wiener-Neustadt, the oft-contemplated translation of his remains never having been carried into effect. It was Innsbruck's fault, as we have seen, that they were not originally laid to rest there, and it is her retribution to have been denied the honour of housing them hitherto. The monument itself is a pile upwards of thirteen feet long and six high, of various coloured marbles, raised on three red marble steps; on the top is a colossal figure, representing the Kaiser dressed in full imperial costume, kneeling, his face being directed towards the altar—a very fine work, cast in bronze by Luigi del Duca, a Sicilian, in 1582. The sides and ends are divided by slender columns into twenty-four fine white marble compartments,[1] setting forth the

[1] For the convenience of the visitor to Innsbruck, but not to interrupt the text, I subjoin here a list of the subjects. (1.) The marriage of Maximilian (then aged eighteen) with Mary of Burgundy at Ghent. (2.) His victory over the French at Guinegate, when he was twenty. (3.) The taking of Arras thirteen years later; not only are the fighting folk and the fortifications in this worthy of special praise, but there is a bit of by-play, the careful finish of which must not be overlooked; and the figure of one woman in particular, who is bringing provisions to the camp, is a masterpiece in itself. (4.) Maximilian is crowned King of the Romans. The scene is the interior of the Cathedral of Aix-la-Chapelle: the Prince is seated on a sort of throne before the altar; the Electors are busied with their hereditary part in the ceremony; the dresses of the courtiers in the crowd, and the ladies high above in their tribune, are a perfect record for the costumier, so minute are they in faithfulness. (5.) The battle of Castel della Pietra, or Stein am Calliano, the landscape background of which is excellent; the Tirolese are seen driving the Venetians with great fury before them over the Etsch (Adige). (6.) Maximilian's entry into Vienna (1490), in course of the contest for the crown of Hungary after the death of Matthias Corvinus; the figure of Maximilian on his prancing horse is drawn with great spirit. (7.) The siege of Stuhlweissenburg, taken by Maximilian the same year; the horses in this tableau deserve particular notice. (8.) The eighth represents an episode which it must have required some courage to record among the acts of so glorious a reign; it shows Maximilian receiving back his daughter Margaret, when, in 1493, Charles VIII. preferred Anne of Brittany to her. The French envoys hand to the Emperor two keys, symbols of the suzerainty of Burgundy and Artois, the price of the double affront of sending back his daughter and depriving him of his bride, for Anne had been betrothed to him. [Margaret, though endowed with the high qualities of her race, was not destined to be fortunate in her married life: her hand was next sought by Ferdinand V. of Spain for his son Don Juan, who died very shortly after the marriage. She was again married, in 1508, to Philibert Duke of Savoy, who died with-

story of his achievements in lace-like relief. If the treatment of the facts is sometimes somewhat legendary,

out children three years later. As Governor of the Netherlands, however, her prudent administration made her very popular.] (9.) Maximilian's campaign against the Turks in Croatia. (10.) The League of Maximilian with Alexander VI., the Doge of Venice, and the Duke of Milan, against Charles VIII. of France; the four potentates stand in a palatial hall joining hands, and the French are seen in the background fleeing in dismay. (11.) The investiture at Worms of Ludovico Sforza with the Duchy of Milan. The portraits of Maximilian are well preserved on each occasion that he is introduced, but in none better than in this one: Maria Bianca is seen seated to the left of the throne, Sforza kneels before them; on the waving standard, which is the token of investiture, the ducal arms are plainly discernible. (12.) The marriage at Brussels, in 1496, of Philip *der Schöne*, Maximilian's son, with Juana of Spain; the Archbishop of Cambrai is officiating, Maximilian stands on the right side of his son: Charles Quint was born of this marriage. (13.) A victorious campaign in Bohemia in 1504. The 14th represents the episodes of the siege of Kufstein, recorded in the second chapter of these Traditions (1504). (15.) The submission of Charles d' Egmont to Maximilian, 1505. The Kaiser sits his horse majestically; the Duke of Gueldres stands with head uncovered; the battered battlements of the city are seen behind them. (16.) The League of Cambrai, 1508. The scene is a handsome tent in the camp near Cambray; Maximilian, Julius II., Charles VIII., and Ferdinand V., are supposed to meet, to unite in league against Venice. (17.) The Siege of Padua, 1509, the first result of this League; the view of Padua in the distance must have required the artist to have visited the place. (18.) The expulsion of the French from Milan, and reinstatement of Ludovico Sforza, 1512. (19.) The second battle of Guinegate: Maximilian fights on horseback; Henry VIII. leads the allied infantry, 1515. (20.) The conjunction of the Imperial and English forces before Terouenne: Maximilian and Henry are both on foot, 1513. (21.) The battle of Vicenza, 1513. (22.) The Siege of Marano, on the Venetian coast. The 23rd represents a noble hall at Vienna, such details as the pictures on the walls not being omitted: Maximilian is treating with Uladisaus, King of Hungary, for the double marriage of their offspring—Anna and Ludwig, children of the latter, with Ferdinand and Maria, grandchildren of the former—an alliance which had its consequence in the subsequent incorporation of Hungary with the Empire. (24.) The defence of Verona by the Imperial forces against the French and Venetians.

the details and accessories are most painstakingly and delicately rendered, great attention having been paid to the faithfulness of the costumes and buildings introduced, and the most exquisite finish lavished on all. They were begun in 1561 by the brothers Bernhard and Arnold Abel, of Cologne, who went in person to Genoa to select the Carrara tablets for their work; but they both died in 1563, having only completed three. Then Alexander Collin of Mechlin took up the work, and with the aid of a large school of artists completed them in all their perfection in three years more. Around it stands a noble guard of ancestors historical and mythological, cast in bronze, of colossal proportions, twenty-eight in number. It is a solemn sight as you enter in the dusk of evening, to see these stern old heroes keeping eternal watch round the tomb of him who has been called 'the last of the Knights,' *der letzte Ritter*. They have not, perhaps, the surpassing merit of the Carrara reliefs, but they are nobly conceived nevertheless. For lightness of poise, combined with excellence of proportion and delicacy of finish, the figure of our own King Arthur commends itself most to my admiration; but that of Theodoric is generally reckoned to bear away the palm from all the rest. They stand in the following order.

Starting on the right side of the nave on entering, we have:

1. Clovis, the first Christian King of France.

2. Philip 'the Handsome,'[1] of the Netherlands, Maximilian's son, reckoned as Philip I. of Spain, though he never reigned there.
3. Rudolf of Hapsburg.
4. Albert II. the Wise, Maximilian's great-grandfather.
5. Theodoric, King of the Ostrogoths. (455–526.)
6. Ernest *der Eiserne*, Duke of Austria and Styria. (1377–1424.)
7. Theodebert, Duke of Burgundy. (640.)
8. King Arthur of England.
9. Sigmund *der Münzreiche*, Count of Tirol. (1427-96.)
10. Maria Bianca Sforza, Maximilian's second wife. (Died 1510.)
11. The Archduchess Margaret, Maximilian's daughter.
12. Cymburgis of Massovica, wife of Ernest *der Eiserne*. (Died 1433.)
13. Charles the Bold, Duke of Burgundy, father of Maximilian's first wife.
14. Philip the Good, father of Charles the Bold. Founder of the Order of the Golden Fleece.

This completes the file on the right side; on our walk back down the other side we come to—

15. Albert II., Duke of Austria, and Emperor of Germany. (1397–1439.)
16. Emperor Frederick I., Maximilian's father. (1415-95.)

[1] Called by the French Philippe 'le Beau,' in distinction from their own 'Philippe le Bel.'

17. St. Leopold, Margrave of Austria; since 1506 the patron saint of Austria. (1073–1136.)
18. Rudolf, Count of Hapsburg, grandfather or uncle of 'Rudolf of Hapsburg.'
19. Leopold III., 'the Pious,' Duke of Austria, Maximilian's great-grandfather; killed at Sempach, 1439.
20. Frederick IV. of Austria, Count of Tirol, surnamed '*mit der leeren Tasche.*'
21. Albert I., D. of Austria, Emperor. (Born 1248; assassinated by his nephew John of Swabia, 1308.)
22. Godfrey de Bouillon, King of Jerusalem in 1099.
23. Elizabeth, wife of the Emperor Albert II., daughter of Sigismund, King of Hungary and Bohemia. (1396–1442.)
24. Mary of Burgundy, Maximilian's first wife. (1457–82.)
25. Eleonora of Portugal, wife of the Emperor Frederick III., Maximilian's mother.
26. Cunigunda, Maximilian's sister, wife of Duke Albert IV. of Bavaria.
27. Ferdinand 'the Catholic.'
28. Johanna, daughter of Ferdinand and Isabella, and wife of Maximilian's son, Philip I. of Spain.

There is a vast difference in the quality both of the design and execution of these statues; the greater number and the more artistic were cast by Gregor Löffler, who established a foundry on purpose at Büchsenhausen; the rest by Stephen and Melchior Godl, and Hanns

Lendenstreich, who worked at Mühlau, a suburb of Innsbruck. All honour is due to them for the production of some of the most remarkable works of their age; but it was some unknown mind, probably that of some humble nameless Franciscan, to whom is due the conception and arrangement of this piece of symbolism. It originally included, besides the statues already enumerated, twenty-three others, of saints, which were to have received a more elevated station, and it is for this reason that they are much smaller in size. They are now placed in the so-called 'Silver Chapel,' and are too frequently overlooked; but it is necessary to take them into account in order worthily to criticize this great monument. They are as follows:—1. St. Adelgunda, daughter of Walbert, Count of Haynault. 2. St. Adelbert, Count of Brabant. 3. St. Doda, wife of St. Arnulf, Duke of the Moselle. 4. St. Hermelinda, daughter of Witger, Count of Brabant. 5. St. Guy, Duke of Lotharingia. 6. St. Simpert, Bishop of Augsburg, son of Charlemagne's sister Symporiana, who rebuilt the monastery of St. Magnus at Füssen. 7. St. Jodok, son of a king of Great Britain; he wears a palmer's dress. 8. St. Landerich, Bishop of Metz, son of St. Vincent, Count of Haynault, and St. Waltruda. 9. St. Clovis. 10. St. Oda, wife of Duke Conrad. 11. St. Pharaild, daughter of Witger, Count of Brabant. 12. St. Reinbert, brother of the last. 13. St. Roland, brother of St. Simpert. 14. St. Stephen, King of Hungary. 15. St. Venantius, martyr, son of Theodoric, Duke of Lotharingia. 16. St. Waltruda, mother of St. Landerich (No. 8). 17. St. Arnulf, husband of St. Doda (No. 3), afterwards Bishop of Metz. 18. St. Chlodulf, son of St. Waltruda (No. 16), also Bishop of Metz. 19. St. Gudula, sister of St. Albert, Count of Brabant. 20. St. Pepin Teuto, Duke of Brabant. 21. St. Trudo, priest, son of St. Adela. 22. St. Vincent, monk. 23. Richard Cœur-de-Lion. A series of men and women, all more or less closely connected with the House of Hapsburg, selected for the alleged holiness of their lives or deeds under one aspect or another. It needs no laboured argument to show the appropriateness of thus representing to the life the solidarity of piety and worth in the great hero's earthly family, though a few words may not be out of place to distinguish the characters allied only or chiefly by the ties of the great family of chivalry. These are—1. King Arthur (No. 8), representative of the mythology of the Round Table. 2. Roland (No. 13 in the series of the saints), representing the myths of the Twelve Peers of France. 3. Theodobert (No. 7), who received a hero's death in the plain of Chalons at the hand of Attila, to be immortalized in the Western Niebelungen Myths. 4. Theodoric (No. 5), celebrated as 'Dietrich von Bern' in the Eastern. 5. Godfrey de Bouillon (No. 22), representing the legendary glory of the Crusades.[1]

[1] This monument earned Ferdinand the title of the Lorenzo de' Medici of Tirol.

The two other statues, of a later date—St. Francis and St. Clare—are by Moll, a native of Innsbruck, who became a sculptor of some note at Vienna. The picture of St. Anthony over the altar of the Confraternity of St. Anthony, on the Epistle side of this church, has a great reputation among the people, because it remained uninjured in a fire which in 1661 burnt down the church of Zirl, where it was originally placed.[1] Five years later it was brought hither for greater honour, and was let into a larger painting by Jele of Vienna, representing a multitude of sick and suffering brought by their friends to pray for healing before it. There is not much else in this church that is noteworthy (besides 'the Silver Chapel,' which belongs to the notice of Ferdinand II.). What there is may be mentioned in a few lines, namely—the *Fürstenchor*, or tribune for the royal family, high up on the right side of the chancel, with the adjoining little chapel and its paintings, and cedar-wood organ, the gift of Julius II. to Ferdinand I.; the quaint old clock; and the memory that Queen Christina of Sweden made her abjuration here 28th October 1655. Her conduct on the occasion was, according to local tradition, most edifying. She was dressed plainly in black silk, 'with no other ornament than a large cross on her breast, with five sparkling diamonds to recall the glorious Wounds

---

[1] St. Anthony being the patron invoked against accidents by fire; also against erisypelas, which in some parts of England even is called 'St. Anthony's fire.'

of the Redeemer. The emphasis with which she repeated the Latin profession of faith after the Papal nuncio did not pass unnoticed. The Ambrosian Hymn was sung at the close of the ceremony, and the church bells and town cannon spoke the congratulations of the Innsbruckers on this and the subsequent days of her stay among them. Among other tokens of gladness, several mystery plays (which are still greatly in vogue in Tirol) were represented. Another public ceremony of her stay was the translation of Kranach's Madonna, the favourite picture of Tirol, brought to it by Leopold V. The original altar-piece of the Hofkirche, by Paul Troger—the Invention of the Cross—was removed by Maria Theresa to Vienna, because the figure of the Empress Helena was counted a striking likeness of herself.

The introduction of the Jesuits into Tirol, and the subsequent building of the Jesuitenkirche in Innsbruck, and the labours of B. Peter Canisius among the people, was also the work of Ferdinand I. The peaceful prosperity which his wise government procured for the country, while wars and religious divisions were distracting the rest of Europe, gave opportunity for the development of its literature and art-culture.[1]

One melancholy event of his reign was the outbreak in its last year, of a terrible epidemic, which committed appalling ravages. All who could, including

---

[1] Weber, *Das Land Tirol*, vol. i. p. 218.

the royal family, escaped to a distance; and those who had been stricken with it were removed to the *Siechenhaus*, and isolated from the rest of the population. As has frequently happened on similar occasions, the dread of the malady operated to deprive the sick of the help of which they stood in need. It was when the plague raged highest, and the majority were most absorbed with the thought of securing their own safety, that a poor woman of the people, named Magaretha Hueber, rising superior to the vulgar terror, took upon herself cheerfully the management of the desolate *Siechenhaus*. The example of her courage was all that was needed to bring out the Christian confidence and charity of the masses; and to her devotion was owing not only the relief of the plague-stricken, but the moral effect of her spirit and energy was also not without its fruit in staying the havoc of the contagion; and she is still remembered by the name of *die fromme Siechen*.

Shortly before his death (which happened in 1564), Ferdinand had his second son, Ferdinand II., publicly acknowledged in the Landtag of Innsbruck, *Landesfürst* of Tirol. His own affection for the country had prevented him from suffering its interests to be ever neglected by the pressure of his vast rule; and now when his great age warned him that he would be able to watch over it no longer, he determined to give it once more the benefit of an independent government.

Ferdinand II. seems to have had all the excellent

administrative qualities of his father in the degree necessary for his restricted sphere of dominion. His disposition for the culture of peaceful arts was promoted by the happiness of his family life. The story of his early love, and his marriage in accordance with the dictates of his heart, in an age when matrimonial alliances were too often dictated by political considerations alone, have made one of the romances dearest to the popular mind. The natural retribution of a disturbance of the regular succession to the throne followed, but with Tirol's usual good fortune the consequences did not prove disastrous, as we shall see later on.

Situated at the distance of a pleasant hour's walk from Innsbruck, and forming an exceedingly picturesque object in the views from it, is Schloss Ambras, in ancient times one of the chief bulwarks of the Innthal. Ferdinand I. bought it of the noble family of Schurfen at the time when he nominated his son to the government of the country, and it always remained Ferdinand II.'s favourite residence. Hither he brought home the beautiful Philippine Welser, whose grace and modesty had won his heart at first sight, as she leant forward from her turret window to cast her flowery greeting at the feet of the Emperor Charles Quint when he came into Augsburg, and the young and handsome prince rode by his side. Philippine had been betrothed by her father to the heir of the Fugger family, the richest and most powerful of Augsburg; but her eyes had met Ferdinand's, and that one glance had revealed to both that their happiness lay in union with each other. Fortunately for Philippine she possessed in her mother a devoted confidant and ally. True, Ferdinand could not rest till he had obtained a stolen interview with her; but the true German woman had confidence in the honour and virtue of the reigning House, and the words Philippine, who was truth itself, reported were those of true love, which knows no shame. Nevertheless, the Fugger was urgent, and old Welser—a sturdy upholder of his family tradition for upright dealing—never, they knew, could be brought to be wanting to his word. The warm love of youth, however, is ever a match for the steady calculation of age. While the fathers Welser and Fugger were counting their money-bags, Ferdinand had devised a plan which easily received the assent of Philippine's affection for him, the rather that her mother, for whom a daughter's happiness stood dearer than any other consideration, gave it her countenance and aid. At an hour agreed, Ferdinand appeared beneath the turret where their happiness was first revealed to them; at a little distance his horses were in waiting. Not an instant had he to wait; Philippine, already fortified by her mother's farewell benediction, joined him ere a pang of misgiving had time to enter his mind, an old and trusted family servant accompanying her. Safely the fugitives reached the chapel, where a friendly priest —Ferdinand's confessor, Johann Cavalleriis—waited to bless the nuptials of the devoted pair, the old servant acting as witness. Old Franz Welser was subsequently induced to give his approval and paternal benediction; and if his burgher pride was wounded by his daughter marrying into a family which might look down upon her connexions, he had the consoling reflection that he was able to give her a dowry which many princes might envy; and also in the discovery of a friendly antiquary, that even his lineage, if not royal, was not either to be despised, for it could be traced up to the same stock which gave Belisarius to the Empire!

Ferdinand's marriage was, I believe, never known to his father; though there are stories of his being won over to forgive it by Philippine's gentle beauty and worth, but these are probably referable to the succeeding Emperor. However this may be, the devoted pair certainly lived for some time in blissful retirement at Ambras; and after his brother, Maximilian II., had acknowledged the legality of Ferdinand's marriage—on the condition that the offspring of it should never claim the rank of Archdukes of Austria—Ambras, which had been their first retreat, was so endeared to them, that they always loved to live there better than anywhere else. There were born to them two sons—Karl, who afterwards became a Cardinal and Bishop of Brixen; and Andreas, Markgrave of Burgau, to whom Ferdinand willed Ambras, on condition that he should maintain its regal beauties, and preserve undiminished the rich stores of books and rare manuscripts, coins, armour, objects of vertù, and curiosities of every sort which it had been the delight of his and Philippine's leisure hours to collect. This testamentary disposition the son judged would be best carried out by selling the place to the Emperor Rudolf II. in 1606; and Ambras has accordingly ever since been reckoned a pleasure-seat of the imperial family. The unfortunate love of centralization, more than the fear of foreign invasion, which was the ostensible pretext, deprived Tirol of these treasures. They were removed to Vienna in 1806, where they may be visited in the Belvedere Palace, the promise of restoring them, often made, not having yet been fulfilled. Among the remnants that are left, are still some tokens of Ferdinand's taste and genius, and some touching memorials of thirty years of happiness purer and truer than had often before been combined with the enjoyment of power. There are some pieces of embroidery, with which Philippine occupied her lonely hours while Ferdinand's public duties obliged him to be away from her, among them a well-executed Crucifixion; and some natural curiosities in the shape of gnarled and twisted roots, needing little effort of the imagination to convert into naturally—perhaps supernaturally—formed crucifixes, and which they had doubtless found pleasure in unearthing in the woods round Ambras. At the time of my visit the private chapel was being very well restored, and some frescoes very fairly executed by Wienhold, a local artist who has studied in Rome. There is still a small collection of armour, and a suit of clothes worn by a giant in the suite of Charles Quint, which would appear to have belonged to a man near eight feet high; also some portraits of the Hapsburg family and other rulers of Tirol; among them Margareta Maultasch, which, if it be faithful, disproves the story deriving her name from the size of her mouth; but of this I shall have occasion to speak later. Inglis mentions that among the relics is a piece of the tree on which Judas hanged himself, but it was not shown to me.

The people, whose own experience fixes the law of suffering in their minds, will have it that these years of tranquil joy were not unalloyed; but that Philippine's mother-in-law embittered them by her jealous bickerings and reproaches, and that these in the end led her to make a sacrifice of her life to the exigencies of her husband's glory. The bath is yet pointed out at Ambras where she is said to have bled herself to death to make way for a consort more conformable to her husband's birth. All, even local, historians, however, are agreed in rejecting this tradition.[1] It has served nevertheless to endear her to the popular mind, for whom she is still a model of domestic virtues no less than a type of beauty. Scarcely is there a house in Tirol that is not adorned by her image. Among other

---

[1] *Zoller Geschichte der Stadt Innsbruck*, p. 272; and *Weicsegger*, vol. vi. p. 61.

traditions of her personal perfections, it is fabled that her skin was so delicate that the colour of the red wine could be seen softly opalized as it passed her slender throat.[1]

[1] I have met the same hyperbole in a piece of homely Spanish poetry.

CHAPTER IX.

NORTH TIROL—THE INNTHAL.

INNSBRUCK (*continued*).

Ora conosce come s'innamora
Lo Ciel del giusto rege, et el sembiante
Del suo fulgore il fa vedere ancora.
DANTE PARADISO, xx. 63.[1]

ANOTHER local tradition of Ambras attaches to a spot where Wallenstein, while a page in the household of Ferdinand and Philippine, fell unharmed from the window of the corridor leading to the dining-hall, making in the terrible moment a secret vow to the Blessed Virgin of his conversion if he escaped with life, which hastened the work begun doubtless by Philippine's devout example and teaching. There is another, again, more marvellous still, and dated from an earlier period, and shortly before the purchase of the castle by the reigning family. It is said that Theophrastus Paracelus, of whom many weird stories are told, was at one time sojourning at Innsbruck—where, another tradition has it, he died—and in the course of his wanderings in search of plants of strange healing powers, came to this outlying and then neglected castle. A peasant woman seeing him pass her cottage weary and

---

[1] 'Now he knows how the just monarch is beloved of Heaven; his beaming countenance yet testifies his joy.

footsore, asked him to come in and rest and taste her freshly-baked cakes, of which the homely odour scented the air. The man of strange science thanked her for her hospitality, and in return touched the tongs upon the hearth with his wonder-working book, and behold the iron was turned into pure gold. The origin of such a legend as this is easy to trace; the book of the *touch* of which such virtue is fabled, plainly represents the learning of the studious savant, which brought him, as well as fame, pecuniary advantage, enabling him to astonish the peasants with payment in the precious metal not often seen by them. But there are many others told of him, the details of which are more complicated, and wander much further from the outline of fact. The way in which he became possessed of his wonder-working power is thus accounted for.[1] One Sunday morning, when he was after his custom wandering in search of plants in a forest on the heights not far from Innsbruck, he heard a voice calling him out of a tree. 'Who are you?' cried Paracelsus. 'I am he whom men call the Evil One,' answered the voice; 'but how wrong they are you shall judge; if you but release me out of this tree you shall see I am not evil at all.' 'How am I to set about it?' asked the clever Doctor. 'Only look straight up the stem of the pine opposite you, and you will see a bung with three crosses on it; all you have to do is to pull it out, and I am free; if you do this I will show you how good I am by giving you the two things you most

[1] Nork, *Mythologie der Volkssagen*, p. 419.

desire, an elixir which shall turn all to gold, and another which shall heal every malady.' Paracelsus, lured by the tempting promise, pulled out the bung, and straightway an ugly black spider crawled out of the hole, and quickly transformed itself into an old man wrapped in a scarlet mantle. The demon kept his word, and gave the Doctor the promised phials, but immediately began threatening the frightful vengeance he would wreak on the exorcist who had confined him in the tree. Paracelsus now blamed himself for his too ready confidence in the character the demon had given himself for goodness, and bethought him of a means of playing on the imp's vanity. 'What a knowing man that same exorcist must be,' said Paracelsus, 'to turn a tall powerful fellow like you into a spider, and then drive you into a tree.' 'Not a bit of it,' replied the imp, piqued, 'he couldn't have done anything of the sort, it was all my own doing.' 'Your own doing!' exclaimed Paracelsus, with a mocking laugh. 'Is that likely? I have heard of people being transformed by some one of greater power than themselves, never by their own.' 'You shall see, though,' said the provoked imp; and with that he quickly resumed the form of a spider, and crawled back into the hole.[1] Paracelsus, it may well be imagined, lost no time in replacing the bung, on which he cut three fresh

---

[1] Exactly the story of the fisherman and the Genius in the copper vessel of the *Arabian Nights*. It is found also in Grimm's story of the Spirit in the bottle, in the Norse tale of the Master Smith; in that of the Lad and the Devil (Dasent); and in the Gaelic tale of the Soldier (Campbell).

crosses to renew the spell; and never can he again be released, for it was agreed never to cut down this forest on account of the protection it afforded to the country against the avalanches.

But, it may be asked, the wonder-working phials once vouchsafed to men, would surely be taken good care of. There is a legend to provide for that too.[1] When the other doctors of Innsbruck found that Paracelsus so far exceeded them in skill, they determined to poison him. Paracelsus had knowledge of their plot by his arts, he knew too that there was only one remedy against the poison they had adopted, and he shut himself up, telling his servant not to disturb him for five days. At the end of the fourth day, however, the curious servant came into his room and broke the spell. Paracelsus had employed a wonder-working spider to draw out the poison, which it would have done in the course of five days. Disturbed on the fourth, Paracelsus knew he must die. Determined that the jealous members of his profession should not profit by their crime, he sent his servant with the two phials and bid him stand in the middle of the Inn-bridge and throw them into the river. Where they fell into the river the water was streaked with molten gold.

It remains to call attention to the splendid and truly Tirolean panoramic view from the pretty terrace of Ambras, with its luxuriant trellis of passion-

---

[1] Von Alpenburg. *Mythen u. Sagen Tirols.*

flower and 'virgin vine.' Overhanging the village of Ambras is the so-called *Tummelplatz*, where in the lifetime of Ferdinand and Philippine, many a gay tournament was held, but since used as a burying-place; first for the military hospital, to which the castle was at one time devoted—and some seven or eight thousand patriots were interred here between 1796 and 1810—and afterwards for those who fell successfully resisting the Italian invasion of 1859.

Whatever was the manner of Philippine's death, it was bitterly lamented by Ferdinand, who found the usual refuge of human grief in raising a splendid monument to her memory, in the so-called *Silberne Kapelle* in the Hofkirche. The chapel had been built by him to satisfy her devotion to the doctrine of the Immaculate Conception; and in her lifetime was so called from the solid silver image of the Blessed Virgin, and the bas-reliefs of the mysteries of the rosary in the same metal over the altar, itself a valuable ebony carving. She had loved to pray there, and it accordingly formed a fitting resting-place for her mortal remains. Her effigy in marble over her altar-shaped tomb is a figure of exceeding beauty, and is ascribed to Alexander Collin; it stands under a marble canopy. The upright slab is of white marble, carved in three compartments; the centre one bearing a modest inscription, and the other two, subjects recording her charity to the living and the dead; the outline of the town of Innsbruck, as it appeared in her day, forms the background. By his desire Ferdinand was buried near her; his monument is similarly sunk in the thickness of the wall, which is adorned with shields carved in relief, bearing the arms of his house painted with their respective tinctures; and on the tomb are marble reliefs, setting forth (after the manner of those on Maximilian's cenotaph) the public acts of his life. This chapel came to be used afterwards for Italian sermons by the consorts of subsequent rulers of Tirol, many of whom were Italians.

In 1572 Innsbruck was visited by a severe shock of earthquake, which overthrew many buildings, and so filled the people with alarm, that temporary wooden huts were built in the open field where they took refuge. Ferdinand and Philippine had recourse to the same means of safety; and while living thus, their only daughter, Anna Eleonora, was born. In thanksgiving for this favour, and for the cessation of the panic, the royal pair vowed a pilgrimage to Seefeld,[1] which they accomplished on foot, accompanied by their sons; above two thousand Innsbruckers following them. The general sentiment of gratitude was further testified by the enactment on the part of Ferdinand, and the glad acceptance on the part of the people, of various rules of devotion, which have gone to form the subsequent habits of the people. Three years of dearth succeeded the earthquake, and were accepted by the

---

[1] See pp. 194, 270, 324–5.

pious ruler and people as a heavenly warning to lead them to increased faith and devotion. Many Lutheran books which had escaped earlier measures against them were spontaneously brought forward and burnt; special devotion to the Blessed Sacrament was promoted, Ferdinand himself setting the example; for whenever he met the Viaticum on the way to the sick, whether he was in a carriage or on horseback, he never failed to alight and kneel upon the ground, whatever might be its condition. This was indeed a special tradition of his house; it is told of Rudolf of Hapsburg, that one day as he was out hunting, a furious storm came on, soon swelling the mountain torrents and sweeping away paths and bridges. On the brink of a raging stream, which there was no means of crossing, stood a priest, weather-bound on his way to carry the last sacrament to a dying parishioner. Rudolf recognised the sound of the bell, and directed his steps by its leading to pay his homage to the '*hochwürdigste Gut.*' He no sooner learned the priest's difficulty than he dismounted, and offered him his own horse. When the priest brought the animal back next day, the pious prince told him he could not think of himself again crossing a horse which had been honoured by having borne his Lord and Redeemer, and begged him to keep it for the future service of religion.

While Philippine's relations never sought to over-step the limits which imperial etiquette had set them, Ferdinand seems to have treated them with kind cordiality. An instance of this was the magnificence with which he celebrated the marriage of her nephew, Johann von Kolourat, with her maid-of-honour, Katarina von Boimont, in 1580: the '*Neustadt*' or principal street afforded space for tournaments and races which lasted many days, and attracted the remaining votaries of chivalry from all parts of Europe. The festivities were closed by a splendid pageant, in which Ferdinand took part as 'Olympian Jove.'

In 1582 Ferdinand married Anna Katharina Gonzaga, daughter of the Duke of Mantua, who was no less pious than Philippine. The marriage was celebrated at Innsbruck with great pomp. She was the first to introduce the Capuchin Order into Germany. Some discussion in the general chapter of the Order preceded the decision which allowed the monks to accept the consequences of being exposed to a colder climate than that to which they had been used. The first stone of their monastery was laid by Ferdinand and Anna Katharina in August 1593, at the intersection of the Universitäts-gasse and the Sill-gasse. Ferdinand died the following year, regretted by all the people, but by none more than by Anna Katharina, who passed the remainder of her days in a convent she had founded at Innsbruck. She died in 1621, and desired the following inscription to be put on her tomb:—'*Miserere mei Domine dum veneris in novissimo die.*'

The warning of the disastrous years 1572-4 was further turned to practical account by Ferdinand in his desire to relieve the distress of the peasants. In the first months of threatening famine he bought with his own means large stores of grain in Hungary and Italy, and opened depôts in various parts of Tirol, where it was sold at a reasonable price. To provide a means of earning money for those who were shut out of their ordinary labour, he laid out or improved some of the most important high roads; he likewise exerted himself in every way to promote the commerce of the country. His reign conferred many other benefits on the people. Many laws were amended and brought in conformity with the altered circumstances of the age; the principle of self-taxation was established, and other measures enacted which it does not belong to my present province to particularise. He introduced also the use of the Gregorian Calendar, and gave great encouragement to the cultivation of letters. It was by his care that the most authentic MSS. of the Nibelungen poems and other examples of early literature were preserved to us.

As Ferdinand had no children by Anna Katharina, and those of Philippine were not allowed to succeed,[1] the rule over Tirol went back at his death to the Emperor Rudolf II., Maximilian's eldest son. In 1602, however, he gave over the government to his brother Maximilian, who is distinguished by the name of the

---

[1] They accepted their position with the usual Tirolese loyalty, and never attempted to found any claims to power on the circumstance of their birth.

T

Deutschmeister. Tirol was again fortunate in her ruler; Maximilian was as pious and prudent a prince as his predecessors. He promoted the educational establishments of the town, and was a zealous opponent of religious differences; he brought in the Order of Servites to oppose the remaining germs of Lutheran teaching; the church and monastery at the end of the Neustadt being built for them by Katharina Maria. There are some pictures in the church by Theophilus Polak, Martin Knoller, Grasmair, and other native artists; and the frescoes on the roof by Schöpf are worth attention. A fanatic named Paul Lederer, one of the very few Tirol has produced, rose in this reign, and carried away about thirty persons to join a kind of sect which he attempted to form; in accordance with the laws of the age, he was tried and executed, after which his followers were no more heard of.

Maximilian was much attached to the Capuchins, and built himself a little hermitage within their precincts, which is still shown, where he spent all the time he could spare in prayer and meditation; following the rule of the monks, rising with them to their night Offices, and employing himself at manual labour in the field and in the workshop like one of them. His cell is paneled with plain wood, the bed and chair are of the most ordinary make, as are the ink-stand and other necessary articles, mostly his own handiwork; it has a window high up in the chancel, whence he could assist at the Offices in the church. The Empress Maria Theresa visited it in 1765, and seating herself in the stiff wooden chair, exclaimed, 'What men our forefathers were!' Another illustrious pilgrim, whose visit is treasured in the memories of the house, was St. Lorenzo of Brindisi, when on his way to found a house of the Order in Austria. The monks begged of him his Hebrew Bible, his walking-stick, and breviary, which are still treasured as relics. All the churches of Innsbruck and many throughout Tirol felt the benefit of Maximilian's devotion to the Church. His spirit was emulated by the townspeople, and when the fatal epidemic of 1611 ceased its ravages, the burghers of Innsbruck built the Dreiheiligkeitskirche[1] for the Jesuits, as a thank-offering that the plague was stayed.

The temporal affairs of Tirol received no less attention from Archduke Maximilian than the spiritual. With the foresight of a true statesman, he discovered the coming troubles of the Thirty Years' War, and resolved that the defences of his country should be in a state to keep the danger at a distance from her borders. The fortified towers, especially those commanding the passes into the country, were all overlooked, and plans of them carefully prepared, all the fortifications being put in repair. The Landwehr, the living bulwarks, the ready defenders of their beloved mountain Vaterland, attracted his still more special attention, and he furnished them

[1] Holy Trinity Church.

with a regulation suited to the needs of the times. He settled also several outstanding disputes with the Venetians, with Count Arco, and with neighbours over the north and west frontiers; and an internal boundary quarrel between the Bishops of Brixen and Trent. The death of Rudolf II., in 1612, had invested him with supreme authority over the country, and simplified his action in all these matters for the benefit of the commonwealth.

Another outburst of pestilence occurred in 1611; the old *Siechen-haus* was not big enough for all the sick, and had no church attached to it. Two Jesuits— the professor of theology at their university, and Kaspar von Köstlan, a native of Brixen—assisted by a lay-brother, devoted themselves to the service of the sick; their example so edified the Innsbruckers, that in their admiration they readily provided the means, at their exhortation, to build a church. Hanns Zimmermann, Dean of the Burgomasters, bound himself by a vow to see to the erection of the building, and from that time it was observed the fury of the pestilence began to diminish. Maximilian bought the neighbouring house and appointed it for the residence of the chaplain of the *Siechen-haus* and the doctors. He gave also the altar-piece by Stötzl, representing the three *Pestschutzheili-*

---

[1] Patron saints against pestilence: viz. SS. Martha (because according to her legend she built a hospital and ended her life tending the sick), Sebastian (because a plague was stayed in Rome at his intercession), and Rocchus (because of the well-known legend of his self-devotion to the plague-stricken).

gen,[1] and another quaint and curious picture of the plague-genius.

Maximilian died in 1618, and a religious vow having kept him unmarried, the government was transferred to Leopold V., Archduke of Styria, again a most exemplary man. His father was Charles II., son of the Emperor Ferdinand I.; he had orginally been devoted to the ecclesiastical state, and nominated Bishop of Strasburg and Passau; but out of regard for the exigencies of the country a dispensation, of which I think history affords only two or three other examples, was granted him from Rome. He married the celebrated Claudia de' Medici, Duchess of Urbino. Though also Governor of the Low Countries, he by no means neglected the affairs of Tirol. Some fresh attempts of Lutherans to interfere with its religious unity, as well as to foment political dissensions, were put down with a resolute hand. Friedrich von Tiefenbach, sometime notorious as a politico-religious leader in Moravia, was discovered in a hiding-place he had selected, in the wild caves at Pfäffers[1] below Chur, and tried and beheaded at Innsbruck in 1621. The selection of Innsbruck for the marriage of the Emperor Ferdinand II. with his second wife Eleonora, daughter of the Duke of Mantua, in 1622, revived the splendours of Maximilian's reign, for the Emperor stayed there some weeks with all his court; the *Landwehr* turned out three thousand strong to form his guard of honour. It was the depth of winter,

[1] Mentioned in the chapter on Vorarlberg, p. 23.

but the bride braved the snow; the Count of Harrach was sent out to meet her on the Brenner Pass with six gilt sledges, and a vast concourse of people. It is recorded that the Emperor wore on the occasion an entirely white suit embroidered with gold and pearls, on his shoulders a short sky-blue cloak lined with cloth of gold, and a diamond chain round his neck. Eleonora, more in accordance with the season, wore a tight-fitting dress of carnation satin embroidered in gold, over it a sable jacket, and a hat with a plume of eagles' feathers. The banquet was entirely served by young Tirolean nobles. The Emperor's present to his bride was a pearl *parure*, costing thirty thousand ducats; and that of the town of Innsbruck a purse of eighteen thousand ducats. Leopold was confirmed by his imperial brother in the government on this occasion. His own marriage was celebrated with scarcely less state than the Emperor's in April 1626, an array of handsome tents being pitched in the meadows of Wilten, where the *Landesschützen* performed many marksmen's feats for the diversion of the company assembled for the ceremonial. This included the Archbishop of Salzburg, who officiated in the Church function, one hundred and fifty counts and barons, and three hundred of noble blood. The visit of the Grand Duke of Tuscany in 1628, and of Ferdinand, King of Hungary and Bohemia, in 1629, were other notable occasions of rejoicing for Innsbruck.

Leopold benefited and adorned the town by the enclosure and planting of the *Hofgarten*, and the bronze equestrian statue of himself, still one of its chief ornaments; but his memory has been more deeply endeared to the people by the present of Kranach's Madonna, which they have copied in almost every church, household, and highway of the country. It is a little picture on panel, very like many of its date, in which the tenderness of devotion beams through and redeems all the stiffness of mannerism; but which we are apt to pass, I had almost said by the dozen, in the various galleries of Europe, with no more than a casual glance. With the Tirolese it was otherwise. Their faith-inspired eyes saw in it a whole revelation of Divine mercy and love; they gazed on the outpouring of maternal fondness and filial confidence in the unutterable communion of the Mother and the Son there portrayed; and deeming that where so much love reigned no petition could be rejected, they believed that answers to the frequent prayers of faith sent up before it were reaped an hundredfold,[1] and the fame of the benefits so derived was symbolized in the title universally given to the picture, of *Mariähülfsbild*.[2] Leopold being in the early part of his reign on a visit to the Elector of Saxony, on occasion of one of his journeys between Tirol and the Low Countries, and being lost in admiration of his collection of pictures at Dresden, received from him the offer of any painting he liked to select.

---

[1] Thirteen volumes were filled with the narrations of such 'answers' received between 1662 and 1665.

[2] Picture of Mary 'Help of Christians'—*Auxilium Christianorum*.

There were many choice specimens, but the devotional conception of this picture carried him away from all the rest, and it became the object of his selection. He never parted from it afterwards, and it accompanied him in all his journeyings. When in Innsbruck, it formed the altar-piece of the *Hofkapelle*, whither the people crowded to kindle their devotion at its focus. After the withdrawal of the allied French, Swedish, and Hessian troops in 1647, the Innsbruckers, in thanksgiving for the success of their prayers before it, built the elegant little circular temple[1] on the left bank of the Inn, still called the Mariähülfskirche, thinking to enshrine it there; but Ferdinand Karl, who had then succeeded to his father Leopold, could not bear to part with it, and gave them a copy instead, by Paul Schor, inserted in a larger picture representing it borne by angels, and the notabilities of Innsbruck kneeling beneath it, the Mariähülfskirche being introduced into the background landscape. However, the number of people who pressed to approach it was so great that he was in a manner constrained to bestow it on the Pfarrkirche only two or three years later, where it now remains; it was translated thither during Queen Christina's visit, as I have mentioned above. It was borne on a car by six white horses, the crowded streets being strewn with flowers. It is a small picture, and has been let into a large canvas painted

---

[1] Inglis says that Schor was the architect of this church, and that he had assisted in building the Vatican.

in Schöpf's best manner, with angels which appear to support it, and beneath St. James, patron of the church, and St. Alexius. A centenary festival was observed in memory of the translation by Maria Theresa in 1750, when all the precious *ex votos*, the thank-offerings for many granted prayers, were exposed to view under the light streaming from a hundred silver candelabra, the air around being perfumed by the flowers of a hundred silver vases. The procession was a splendid pageant, in which no expense seems to have been spared, the great Empress herself, accompanied by her son, afterwards Joseph II., heading it. This was repeated—in a manner corresponding with the diminished magnificence of the age—in 1850, the Emperor Ferdinand I., the Empress Anna, and other members of the Imperial family, taking their part in it.[1]

The only remaining act of Leopold's reign which calls for mention in connexion with Innsbruck, was the erection of the monument to Maximilian the Deutschmeister, in the Pfarrkirche, almost the only one that was spared when the church was rebuilt after the earthquakes of 1667 and 1689, the others having been

---

[1] It is painted on panel, thirty inches by twenty-one; the figure of our Lady is three quarter-length, but appears to be sitting, as the foot of the Divine Infant seems to rest upon her knee. The tradition concerning it is, that it represents an episode of the Flight into Egypt, when, as the Holy Family rested under a palm-grove, they were overtaken by a band of robbers, headed by S. Demas, the (subsequently) penitent thief. The Holy Child is indeed represented *clinging* to His Mother—not as in fear, or even as if need were to suggest courage to her, but simply as if an attack sustained in common impelled a closer union of affection.

ruthlessly used—the headstones in building up the walls, the bronze ones in the bell-castings.

Leopold's son, Ferdinand Karl, being under age at the time of his death, in 1632, he was succeeded by his widow, Claudia de' Medici, as regent. The troubles of the Thirty Years' War, in which Leopold like other German princes had had his chequered share, were yet raging. Claudia was equal to the exigencies of her time and country. She continued the measures of Maximilian the *Deutschmeister* for perfecting the defences of the country, and particularly all its inlets; and she encouraged the patriotic instincts of the people by constantly presiding at their shooting-practice. The Swedish forces, after taking Constance, advanced as far as the Valtelin, and Tirol was threatened with invasion on both sides at once. By her skilful measures, at every rumour of an inroad, the mountains bristled with the unerring marksmen of Tirol, securely stationed at their posts inaccessible to lowlanders. Nothing was spared to keep up the vigilance and spirit of the true-hearted peasants. By this constant watchfulness she saved the country from the horrors of war, in which almost the whole of the German Empire was at that time involved. During all this time she was also developing the internal resources, and consolidating the administration of the country. Two misfortunes, however, visited Inusbruck during her reign: a terrible pestilence, and a destructive fire in which the Burg suffered severely, the beautiful chapel of

Ferdinand II. being consumed, and the body of Leopold, her husband, which was lying there at the time, rescued with difficulty. After this, Claudia spent some little time at Botzen, and also visited Florence. It may be questioned whether the introduction of the numerous Italians about her court was altogether for the benefit of Tirol. They brought with them certain ways and principles which were not altogether in accordance with the German character; and we have seen the effect of the jealousies of race in the tragic fate of her chancellor Biener.[1]

Ferdinand Karl having attained his majority in 1646, Claudia withdrew from public affairs, and died only two years later. In his reign the introduction of the Italian element at court was apparent in the greater luxury of its arrangements, and in the greater cultivation of histrionic and musical diversions. The establishment of the theatre in Innsbruck is due to him. The marriage of his two sisters, Maria Leopoldina and Isabella Clara, and the frequent interchange of visits between him and the princes of Italy, further enlivened Innsbruck. The visit of Queen Christina,[2] of which I have already said enough for my limits, also took place in his reign (1655). Nor did Ferdinand Karl give himself up to amusement to the neglect of business, or of

---

[1] See pp. 123-4.

[2] She was on her way to Rome, where she spent the rest of her life. Alexander VII. commissioned Bernini to rebuild the Porta del Popolo, and adorned it with its inscription, *Felici, faustoque ingressui*, in honour of her entry.

more manly pleasures. He maintained all his mother's measures for the encouragement of the *Schiebenschiessen*, and had the satisfaction of seeing the departure of the enemy's army from his borders, which was celebrated by the building of Mariähülfskirche.[1] To his love of the national sport of chamois-hunting his death has to be ascribed; for the neglect of an attack of illness while out on a mountain expedition near Kaltern after the wild game, gave it a hold on his constitution, which placed him beyond recovery. His death occurred in 1660, at the early age of thirty-four; he left no heir.

He was succeeded by his only brother, Sigmund Franz, Bishop of Gurk, Augsburg, and Trent, who seems to have inherited all his mother's finer qualities without sharing her Italianizing tendencies. With a perhaps too sudden sternness, he purged the court and government of all foreign admixture, and reduced the sumptuous suite of his brother to dimensions dictated by usefulness alone. However popular this may have made him with the German population, the ousted Italians were furious; and his sudden death—which occurred while, after the pattern of his father, applying for a dispensation to marry, in 1665—was by the Germans ascribed to secret poisoning; his Tuscan physician Agricola having, it is alleged, been bribed to perpetrate the misdeed.

Tirol now once more reverted to the Empire. Though Leopold I. came to Innsbruck to receive the

[1] See p. 280.

homage of the people on his accession, and a gay ceremonial ensued, yet it lost much of its importance by having no longer a resident court. While there, however, Leopold had seen the beautiful daughter of Ferdinand Karl's widow, Claudia Felicità, who made such an impression upon him, that he married her on the death of his first wife. The ceremony was performed in Innsbruck by proxy only ; but the dowager-archduchess provided great fêtes, in which the city readily concurred, and gave the bride thirty thousand *gulden* for her wedding present. Claudia Felicità, in her state at Vienna, did not forget the good town of Innsbruck ; and by her interest with her husband, Tirol received a Statthalter in the person of Charles Duke of Lotharingia, husband of his sister Eleonora Maria, widow of the King of Poland. Charles took up his residence at Innsbruck ; and though he was often absent with the army, the presence of his family revived the gaiety of the town ; still it was not like the old days of the court. Charles, however, who had been originally educated for the ecclesiastical state, was a sovereign of unexceptionable principles and sound judgment ; and he did many things for the benefit of Tirol, particularly in developing its educational establishments. He raised the Jesuit gymnasium of Innsbruck to the character of a university ; and the privileges with which he endowed it, added to the salubrity of the situation, attracted alumni from far and near, who amounted to near a thousand in number.

Nothing of note occurred in Tirol till 1703—the Duke of Lotharingia had died in 1696—which is a memorable year. The war of the Spanish Succession, at that time, found Maximilian, Elector of Bavaria, and some of the Italian princes, allied with France against Austria—thus there were antagonists of Austria on both sides of Tirol; nevertheless, no attack on it seems to have been apprehended; and thus, when a plan was concerted for entering Austria by Carinthia (the actual boundaries against Bavaria being too well defended to invite an entrance that way), and it was arranged that the Bavarian and Italian allies should assist the French in overrunning Tirol, everyone was taken by surprise. Maximilian easily overcame the small frontier garrison. At Kufstein he met a momentary check, but an accident put the fortress in his power. Possessed of this base of operations, he was not long in reducing the forts of Rottenburg Scharnitz, and Ehrenberg, and possessing himself of Hall and Innsbruck. He now reckoned the country his, and that it only remained to send news of his success to Vendôme, who had taken Wälsch-Tirol similarly by surprise and advanced as far as Trent, in order to carry out their concerted inroad through the Pusterthal. So sure of his victory was he, that he ordered the Te Deum to be sung in all the churches of Innsbruck.

In the meantime the Tirolese had recovered from their surprise, and had taken measures for disconcerting and routing the invaders; the storm-bells and the *Kreidenfeuer*[1] rallied every man capable of bearing arms, to the defence of his country. The main road over the Brenner was quickly invested by the native sharp-shooters; there was no chance of passing *that* way. Maximilian thought to elude the vigilance of the people by sending his men round by Oberinnthal and the Finstermünz. The party trusted with this mission were commanded by a Bavarian and a French officer. They reached Landeck in safety, but all around them the sturdy Tirolese were determining their destruction. Martin Sterzinger, *Pfleger* or Judge, of Landeck, summoned the *Landsturm* of the neighbouring districts, and arranged the plan of operation. The enemy were suffered to advance on their way unhindered along the steep path, where the rocky sides of the Inn close in and form the terrible gorge which is traversed by the Pontlatzerbrücke; but when they arrived, no bridge was there! The mountaineers had been out in the night and cut it down. Beyond this point the steep side afforded no footing on the right bank, no means remained of crossing over to the left! The remnants of the bridge betrayed what had befallen, and quickly the command was given to turn back; in the panic of the moment many lost their footing, and rolled into the rapid river beneath. For those even who retained their composure no return was possible; the heights above were peopled with the ready Tirolese,

[1] *Kreidenfeuer*— alarm fires, from *Krei*, a cry.

burning to defend their country. Down came their shots like hail, each ball piercing its man; those who had no arms dashed down stones upon the foe. Only a handful escaped, but at Landeck these were taken prisoners; and there was not *one* even to carry the news to Maximilian. This famous success is still celebrated every year on the 1st of July by a solemn procession.

Maximilian and Vendôme remained perplexed at hearing nothing from each other, and without means of communication; in vain they sent out scouts; money could not buy information from the patriotic Tirolese. Meantime, danger was thickening round each; the *Landsturm* was out, and every height was beset with agile climbers, armed with their unerring carbines, and with masses of rock to hurl down on the enemy who ventured along the road beneath them. The Bavarian and French leaders in the north and in the south only perceived how critical was their situation just in time to escape from it, and the waste and havoc they had made during their brief incursion was recompensed by the numbers lost in their retreat. The Bavarians held Kufstein for some time longer, but their precipitate withdrawal from all the rest of the country earned for the campaign, in the mouths of the Tirolese, the nickname of the *Baierische-Rumpel*. While brave arms had been defending the mountain passes, brave hearts of those whose arms were nerved only for being lifted up in prayer, not for war, were day by day earnestly interceding in the churches for the deliverance of their husbands, fathers, and brothers ; and when, on the 26th of July, the land was found free of the foe, it was gratefully remembered that it was S. Anne's Day, and the so-called *Annensäule*, which adorns the *Neustadt*—the principal thoroughfare of Innsbruck—was erected in commemoration.

It is composed of the marbles of the country; the lower part red, the column white, the effigy of the Immaculate Conception, which surmounts it and the surrounding rays, in gilt bronze. Round the base stand St. Vigilius and St. Cassian (two apostles of Tirol), and St. Anne and St. George; about them float angels, in the breezy style of the period. The monument was solemnly inaugurated on S. Anne's Day, 1706; and every year on that day a procession winds round it from the parish church, singing hymns of thanksgiving; and an altar, gaily dressed with fresh flowers, stands before it for eight days under the open sky.

Leopold I. died in 1705, and was succeeded by his son, Joseph I., who reigned only six years. Charles VI., Leopold's younger son, followed, who appointed Karl Philipp, Palsgrave of Neuburg, Governor of Tirol. He was another pious ruler, and much beloved by the people; his memory being the more endeared to them, that he was their last independent prince. His reign benefited Innsbruck by the erection of the handsome *Landhaus* and the *Gymnasium*, and also by the extensive restoration of the Pfarrkirche. This occupied the site of the little chapel, the accorded privilege to which of hearing in it masses of obligation forms the earliest record of Innsbruck's history. It had grown with the growth of the town, and had been added to by various sovereigns, and we have seen it gifted with Kranach's *Mariähilf*. The earthquakes of 1667 and 1689 had left it so dilapidated, however, that Karl Philipp resolved to rebuild it on a much larger plan. He laid the first stone on May 12, 1717, in presence of his brother, the Bishop of Augsburg, and it was consecrated in 1724. It has the costliness and the vices of its date; its overloaded stucco ornaments are redeemed by the lavish use of the beautiful marbles of the country; the quarrying and fashioning these marbles occupied a hundred workmen, without counting labourers and apprentices, for the whole time during which the church was building. The frescoes setting forth the wonder-working patronage of St. James, on the roof and cupola, are by Kosmas Damian Asam, whose pencil, and that of his two sons, Kosmas and Egid, were entirely devoted to the decoration of churches and religious houses. There is a tradition, that as the fervent painter was putting the finishing touches to the figure of the saint, as he appears, mounted on his spirited charger as the patron of Compostella, in the cupola, he stepped back to see the effect of his work. Forgetting in his zeal the narrowness of the platform on which he stood, he would inevitably have been precipitated on to the pavement below, but that the strong arm of the saint he had been painting so lovingly, detached itself from the wall, and saved his client from the terrible fate! [1] Other works of this reign were the *Strafarbeitshaus*, a great improvement on the former prison; and the church of St. John Nepomuk, in the Innrain, then a new and fashionable street. The canonization of the great martyr to the seal of Confession took place in 1730. Though properly a Bohemian saint, his memory is so beloved all through southern Germany, that all its divisions seem to lay a patriotic claim to him. His canonization was celebrated by a solemn function in the Pfarrkirche, lasting eight days; and the people were so stirred up to fervour by its observance, that they subscribed for the building of a church in his honour, the governor taking the lead in promoting it.

Maria Theresa succeeded her father, Charles VI., in 1742. She seems to have known how to attend to the affairs of every part of the Empire alike; and thus, while the whole country felt the benefit of her wise provisions, all the former splendours of the Tirolean capital revived. Maria Theresa frequently took up her residence at Innsbruck; and while benefiting trade by her expenditure, and by that of the visitors whom her court attracted, she set at the same time an edifying

---

[1] A leading spiritualist, who has also a prominent position in the literary world, tells the story that one day he had missed his footing in going downstairs, and was within an ace of making as fatal a fall as Professor Phillips, when he distinctly felt himself seized, supported, and saved by an invisible hand. The analogy between the two convictions is curious.

example of piety and a well-regulated life. Her associations with Innsbruck were nevertheless overshadowed by sad events more than once, though this does not appear to have diminished her affection for the place.

When Marshal Daun took a whole division of the Prussian army captive at Maxen in 1758, the officers, nine in number, were sent to Innsbruck for safe custody. Here they remained till the close of the war, five years later. This, and the furnishing some of its famous sharpshooters to the Austrian contingent, was the only contact Tirol had with the Seven Years' War. Two years after (1765) Maria Theresa arranged that the marriage of her son (afterwards Leopold II.) with Maria Luisa, daughter of Charles III. of Spain, should take place there. The townspeople, sensible of the honour conferred on them, responded to it by adorning the city with the most festive display; not only with gay banners and hangings, but by improving the façades of their houses, and the roads and bridges, and erecting a triumphal arch of unusual solidity at the end of the Neustadt nearest Wilten, being that by which the royal pair would pass on their way from Italy; for Leopold was then Grand Duke of Tuscany. The theatre and public buildings were likewise put in order. Maria Theresa, with her husband Francis I., and all the Imperial family, arrived in Innsbruck on July 15, attracting a larger assemblage of great people than had been seen there even in its palmiest days. Banquets and gay doings filled up the interval till August 5, when Leopold and Maria Luisa made their entrance with unexampled pomp. The marriage was celebrated in the *Pfarrkirche* by Prince Clement of Saxony, Bishop of Ratisbon, assisted by seven other bishops. Balls, operas, banquets, illuminations, and the national *Freischiessen*, followed. But during all these fêtes, an unseasonable gloom, which is popularly supposed to bode evil, overclouded the August sky, usually so clear and brilliant in Innsbruck. On the 18th, a grand opera was given to conclude the festivities; on his way back from it Francis I. was seized with a fit, and died in the course of the night in the arms of his son, afterwards Joseph II.

Though Maria Theresa's master mind had caused her to take the lead in all public matters, she was devotedly attached to her husband, and this sudden blow was severely felt by her. She could not bear that the room in which he expired should ever be again used for secular purposes, and had it converted into a costly chapel; at the same time she made great improvements and additions to the rest of the Burg. She always wore mourning to the end of her life, and always, when state affairs permitted, passed the eighteenth day of every month in prayer and retirement. A remarkable monument remains of both the affection and public spirit of this talented princess. Driving out to the Abbey of Wilten in one of the early days of mourning, while some of the tokens of the rejoicing, so unexpectedly turned into lamentation, were still unremoved, the sight of the handsome triumphal arch reminded her of a resolution suggested by Francis I. to replace it by one of similar design in more permanent materials. Her first impulse was to reject the thought as a too painful reminder of the past ; but reflection on the promised benefit to the town prevailed over personal feelings, and she gave orders for the execution of the work ; but to make it a fitting memorial of the occasion, she ordered that while the side facing the road from Italy should be a *Triumphpforte,* and recall by its bas-reliefs the glad occasion which caused its erection, the side facing the town should be a *Trauerpforte,* and set forth the melancholy conclusion of the same. The whole was executed by Tirolean artists, and of Tirolean marbles. She founded also a *Damenstift,* for the maintenance of twelve poor ladies of noble birth, who, without taking vows, bound themselves to wear mourning and pray for the soul of Francis I. and those of his house. Another great work of Maria Theresa was the development she gave to the University of Innsbruck.

After her death, which took place in 1780, Joseph II., freed from the restraints of her influence, gave full scope to his plans for meddling with ecclesiastical affairs, for which his intercourse with Russia had perhaps given him a taste. Pius VI. did not spare himself a journey to Vienna, to exert the effect of his personal influence with the Emperor, who it would seem did not pay much heed to his advice, and so disaffected his people by his injudicious innovations, that at the time of his death the whole empire, which the skill of Maria Theresa had consolidated, was in a state of complete disorganization.¹ Though increased by his ill-gotten share of Poland, he lost the Low Countries, and Hungary was so disaffected, that had he not been removed by the hand of death (1790), it is not improbable it would have thrown off its allegiance also. Leopold II., his brother, who only reigned two years, saved the empire from dissolution by prudent concessions, by rescinding many of Joseph's hasty measures, and abandoning his policy of centralization.

One religious house which Joseph II. did not suppress was the *Damenstift* of Innsbruck, of which his sister, the Archduchess Maria Elizabeth, undertook the government in 1781; and during the remainder of her life held a sort of court there which was greatly for the benefit of the city. Pius VI. visited her on his way back from Vienna on the evening of May 7, 1782. The whole town was illuminated, and all the religious in the town went out to meet him, followed by the whole body of the people. Late as was the hour (a quarter to ten, says a precise chronicle) he had no sooner reached the apartment prepared for him in the Burg, than he admitted whole crowds to audience, and the enthusiasm with which the religious Tirolese thronged round him surpasses words. Many, possessed with a sense of the honour of having the vicar of

---

¹ Consult Cesare Cantù *Storia Universale*, § xvii. cap. 21.

Christ in their very midst, remained all night in the surrounding *Rennplatz*, as it were on guard round his abode. In the morning, after hearing mass, he imparted the Apostolic Benediction from the balcony of the Burg, and proceeded on his way over the Brenner.

Leopold II. had not been three months on the throne before he came to Innsbruck to receive the homage of his loyal Tirolese, who took this opportunity of winning from him the abrogation of many *Josephinischen* measures, particularly that reducing their University to a mere Lyceum. He was succeeded in 1792 by his son, Francis II.; but the mighty storm of the French Revolution was threatening, and absorbed all his attention with the preservation of his empire, and the defence of Tirol seems to have been overlooked. Year by year danger gathered round the outskirts of her mountain fastnesses. Whole hosts were engaged all around; yet there were but a handful, five thousand at most, of Austrian troops stationed within her frontier. The importance of obtaining the command of such a base of operations, which would at once have afforded a key to Italy and Austria, did not escape Bonaparte. Joubert was sent with fifteen thousand men to gain possession of the country, and advanced as far as Sterzing. Innsbruck was thrown into a complete panic, and I am sorry to have to record that the Archduchess Maria Elizabeth took her flight. The Austrian Generals, Kerpen and Laudon, did not deem it prudent, with their small contingent, to engage the French army.

Nevertheless, the Tirolese, instead of being disheartened at this pusillanimity, with their wonted spirit rose as one man; a decisive battle was fought at Spinges, a hamlet near Sterzing, where a village girl fought so bravely, and urged the men on to the defence of their country so generously, that though her name is lost, her courage won her a local reputation as lasting as that of Joan of Arc or the 'Maid of Zaragoza,' under the title of *Das Mädchen von Spinges*.[1] Driven out hence, the French troops made the best of their way to join the main army in Carinthia. After this the enemy left Tirol at peace for some years, with the exception of one or two border inroads, which were resolutely repulsed. One of these is so characteristic of

---

[1] Since writing the above, I have been assured by one who has frequently conversed with her, that the concealment of her name arose from her own modesty; it was Katharina Lanz. To avoid public notice, she went to live at a distance, and up to the time of her death in 1854, bore an exemplary character, living as housekeeper to the priest serving the mountain church of S. Vigilius, near Rost, the highest inhabited point of the Enneberg. When induced to speak of her exploits, she always made a point of observing that, though she brandished her hay-fork, she neither actually killed or wounded anyone. She had heard that the French soldiers were nothing loth to desecrate sacred places, and she stationed herself in the church porch, determined to prevent their entrance; the churchyard had become the citadel of the villagers. From her post of observation she saw with dismay that her people were giving way. It was then she rushed out and rallied them; in her impetuosity she was *very near* running her hay-fork through a French soldier, but she was saved from the deed by her landlord, who, encouraged by her ardour, struck him down, pushing her aside. The success of her sally and her subsequent disappearance cast a halo of mystery round her story, and many were inclined to believe the whole affair was a heavenly apparition.

the religious customs of Tirol, that, though not strictly belonging to the history of Innsbruck, I cannot forbear mentioning it. The French, under Massena, had in 1799 been twice repulsed from Feldkirch with great loss. Divisions which had never known a reverse were decimated and routed by the practised guns of the mountaineers. Thinking their victory assured, the peasants, after the manner of volunteer troops, had dispersed but too soon, to return to their flocks and tillage. Warily perceiving his advantage, Massena led his troops back over the border silently by night, intending in the morning to take the unsuspecting town by storm—a plan which did not seem to have a chance of failure. But it happened to be Holy Saturday. Suddenly, just as he was about to give the order for the attack, the bells of all the churches far and near, which had been so still during the preceding days, burst all together upon his ear with the jubilant *Auferstehungsfeier*.[1] General and troops, alike unfamiliar with religious times and seasons, took the sound for the alarm bells calling out the *Landsturm*. In the belief that they were betrayed, a precipitate retreat was ordered. But the night no longer covered the march; and the peasants, who were gathered in their villages for the Offices of the Church, were quickly collected for the pursuit. This abortive expedition cost the French army three thousand men.

---

[1] Celebration of the Resurrection.

In the meantime the Archduchess had returned to Innsbruck, and all went on upon its old footing, as if there were no enemy to fear. So little was another disturbance expected, that the Archduchess devoted herself to the promotion of local improvements, including that of the *Gottesacker*. This is one of the favourite Sunday afternoon resorts of the Innsbruckers, and is well worthy of a visit. The site was first destined for the purpose by the Emperor Maximilian. It was gifted with all the indulgences accorded to the Campo Santo of Rome by the Pope, and in token of the same some earth from San Lorenzo *fuor le mura* was brought hither at the time of its consecration by the Bishop of Brixen in 1510. It has, according to the frequent German arrangement, an upper and a lower chapel; the former, dedicated to S. Anne; the latter, as usual, to S. Michael, though the people commonly call it *die Veitskapelle*, on account of some cures of S. Vitus' dance wrought here. The arcades which now surround the cemetery were the result of the introduction of Italian customs later in the sixteenth century. Some of the oldest and noblest names of Tirol are to be found upon the monuments here, some of which cannot fail to attract attention. The bas-reliefs sculptured by Collin for that of the Hohenhauser family, and those he prepared for his own, may be reckoned among his masterpieces. Some which are adorned with paintings would be very interesting if the weather had spared them more. The

Archduchess had prepared her own resting-place here also, but was not destined to occupy it. The disastrous defeat of Austerlitz filled her with alarm, and she once more fled from Innsbruck, this time not to return.

This was the year 1805, and a sad one it was for Tirol. The treaty of Pressburg had given Tirol to Bavaria, and Bavaria and Tirol had never in any age been able to understand each other. Willingly would the Tirolese have opposed their entrance; but the Bavarians, who knew every pass as well as themselves, were enabled to pour in the allied troops under Marshal Ney in such force, that they were beyond their power to resist. The fortresses near the Bavarian frontier were razed, and Innsbruck occupied. On February 11, 1806, Marshal Ney left, and the town was formally delivered over to Bavarian rule. The most unpopular changes of government were adopted, particularly in ecclesiastical matters and in forcing the peasants into the army; the University also was once more made into a Lyceum. But the Landsturm was not idle, and the Archduke Johann, Leopold's brother, came into Tirol to encourage them. Maturing their plans in secret, the patriots, under Andreas Hofer, who had been to Vienna in January to declare his plans and get them confirmed by his government, and Speckbacher, broke into Innsbruck on April 13, 1809, where the townspeople received them with loud acclamations; and after a desperate and celebrated conflict at Berg Isel, succeeded in completely ridding it of the invaders.

The Bavarian arms on the Landhaus were shattered to atoms, and when the Eagle replaced them, the people climbed the ladders to kiss it. This was the first great act of the *Befreiungskämpfe* which have made 'the year Nine' memorable in the annals of Tirol, and, I may say of Europe, for it was one of the noblest struggles of determined patriotism those annals have to boast, and at the same time the most successful effort of volunteer arms. Hofer accepted the title of *Schützenkommandant*, and was lodged in the imperial *Burg*, while his peasant neighbours took the office of guards; but he altered nothing of his simple habits, nor his national costume. His frugal expenses amounted to forty-five kreuzers a day, and he lost no opportunity of expressing that he did nothing on his own account, but all in the name of the Emperor. On May 19 the Bavarians laid siege to the town; but the defenders of the country, supported by a few regular Austrian troops, obliged them by the end of a fortnight to decamp. On June 30 they returned with a force of twenty-four thousand men: but other feats of arms of the patriots in all parts of Tirol showed that its people were unconquerable, and for the third time Hofer took possession of Innsbruck. In the meantime, however, the Peace of Schönbrunn, of October 25, had nullified their achievements, though the memory of their bravery could never be blotted out, and always asserted its power. Nor could the brave people, even when bidden by the Emperor himself to desist, believe that his orders were otherwise than wrung from him, nor could their loyalty be quenched. Hofer's stern sense of subordination made him advise abstention from further strife, but the more ardent patriots refused to listen, and ended by leading him to join them. A desultory warfare was now kept up, with no very effectual result, but yet with a spirit and determination which convinced the Bavarians that they could never subdue such a people, and predisposed them to consent to the evacuation of their country in 1814; for they saw that

> Freedom from every hut
> Sent down a separate root,
> And when base swords her branches cut,
> With tenfold might they shoot.

In the meantime a terrible wrong had been committed; the French, knowing the value of Hofer's influence in encouraging the country-people against them, set a price on his head sufficient to tempt a traitor to make know his hiding-place. He was taken, and thrown into prison at the *Porta Molina* at Mantua. Tried in a council of war, several voices were raised in honour of his bravery and patriotism; a small majority, however, had the cowardice to condemn him to death. He received the news of the sentence with the firmness which might have been expected of him, the only favour he condescended to ask being the spiritual assistance of a priest. Provost Manifesti was sent to him, and remained with him to the end. An offer was made him of saving his life by entering the French service, but he indignantly refused to join the enemies of his country. To Provost Manifesti he committed all he possessed, to be expended in the relief of his fellow-countrymen who were prisoners. He spent the early hours of the morning of the day on which he was to die, after mass, in writing his farewell to his wife, bidding her not to give way to grief, and to his other relations and friends, in which latter category was comprehended the population of the whole Passeyerthal, not to say all Tirol; recommending himself to their prayers, and begging that his name might be given out, and the suffrages of the faithful asked for him, in the village church where he had so often knelt in years of peace. He was forbidden to address his fellow-prisoners. He bore a crucifix, wreathed in flowers, in his hand as he walked to the place of execution, which he was observed repeatedly to kiss. There he took a little silver crucifix from his neck, a memorial of his first Communion, and gave it to Provost Manifesti. He refused to kneel, or to have his eyes bandaged, but stood without flinching to receive the fire of his executioners. His signal to them was first a brief prayer; then a fervently uttered 'Hoch lebe Kaiser Franz!' and then the firm command, 'Fire home!' His courage, however, unmanned the soldiers; ashamed of their task, they durst not take secure aim, and it took thirteen shots to send the undaunted soul of the peasant hero to its rest. It was February 20, 1810; he was only forty-five. The traditions of his courage and endurance, his probity and steadfastness, are manifold; but in connexion with Innsbruck we have only to speak of his brief administration there, which was untarnished by a single unworthy deed, a single act of severity towards prisoners of war, of whom he had numbers in his power who had dealt cruel havoc on his beloved valleys.

The Emperor for whom he had fought so nobly returned to Innsbruck, to receive the homage of the Tirolese, on May 28, 1816, amid the loud rejoicings of the people, preceded by a solemn service of thanksgiving in the *Pfarrkirche*. Illuminations and fêtes followed till June 5, when the ceremony was wound up by a grand shooting-match, at which the Emperor presided and many prizes were distributed. The number who contended was 3,678, and 2,137 of them made the bull's-eye; among them were old men over eighty and boys of thirteen and fourteen.

The claims of Hofer on his country's remembrance were not forgotten when she once more had leisure for works of peace. His precious remains, which had been carefully interred by the priest who consoled his last moments at Mantua, were brought to Innsbruck in 1823, and laid temporarily in the *Servitenkloster*. On February 21 they were borne in solemn procession by six of his brothers in arms, all the clergy and people following. The Abbot of Wilten sang the requiem office. The Emperor ordered the conspicuous and appropriate monument to mark the spot where they laid him, which is one of the chief ornaments of the *Hofkirche.* The pedestal bears the inscription—

<blockquote>Seinen in den Befreiungskämpfen gefallenen Söhnen das dankbare Vaterland,</blockquote>

and the sarcophagus the words—

<blockquote>Absorbta est mors in victoria.</blockquote>

Tirol had no reason to regret the restoration of the dynasty for which she had suffered so much. Most of her ancient privileges were restored to her, and in 1826 Innsbruck again received the honour of a University, and many useful institutions were founded. Francis came to Innsbruck again this year, and while there, received the visit of the Emperor of Russia and the King of Prussia. Another shooting-match was held before them, at which the precision of the Tirolese received much praise; and again for a short time in 1835. The Archduke John, who came in 1835 to live in Tirol, was received with great enthusiasm; his hardy feats of mountain climbing, and hearty accessible character, endearing him to all the people.

The troubles of 1848 gave the Tirolese again an opportunity of showing that their ancient loyalty was undiminished. The Emperor Ferdinand, driven out of his capital, found that he had not reckoned wrongly in counting on a secure refuge in Tirol. It was the evening of May 16 that the Imperial pair came as fugitives to Innsbruck. Though there was hardly time to announce their advent before their arrival, the people went out to meet them, took their horses from the carriage, and themselves drew it into the town; and all the time they remained the townspeople and *Landes-schüten* mounted guard round the Burg. More than this, the Tirolese *Kaiser-Jäger-Regiment* volunteered for service against the insurgents, and fought with such determination that Marshal Radetsky pronounced that every man of them was a hero. With equal stout-heartedness the *Landesschützen* repelled the attempted Italian invasion at several points of the south-western frontier, and kept the enemy at bay till the imperial troops could arrive. These services were renewed with equal fidelity the next year. A tablet recording the bravery of those who fell in this campaign—one of the officers engaged being Hofer's grandson—is let into the wall of the Hofkirche opposite Hofer's monument.

It was this Emperor from whom the name of Ferdinandeum was given to the Museum, but it was rather out of compliment, and while he was yet Crown-Prince, than in memory of any signal co-operation on his part. It was projected in 1820 by Count Von Chotek, then Governor of Tirol. It comprises an association for the promotion of the study of the arts and sciences. The Museum contains several early illuminated MSS., in the production of which the Carthusians of Schnals and the Dominicans of Botzen acquired a singular pre-eminence. At a time when the nobles of other countries were occupied with far less enlightened pursuits, the peaceful condition of Tirol enabled its nobles, such as the *Edelherrn* of Monlan, Annaberg, Dornsberg, Runglstein, and others, to keep in their employment secretaries, copyists, and chaplains, busied in transcribing; and often sent them into other countries to make copies of famous works to enrich their collections. It has also some of the first works produced from the printing-press of Schwatz already mentioned. This press was removed to Innsbruck in 1529; Trent set one up about the same time. In the lower rooms of the Ferdinandeum is a collection of paintings by Tirolean artists, and specimens of the marbles, minerals, and other natural productions of the country. The great variation in the elevation of the soil affords a vast range to the vegetable kingdom, so that it can boast of giving a home to plants like the tobacco, which only germinates at a temperature of seventy degrees, and the edelweiss, which only blossoms under the snow. There is also a small collection of Roman and earlier antiquities, dug up at various times in different parts of Tirol, and specimens of native industries. Among the most singular items are some paintings on cobweb, of which one family has possessed the secret for generations, specimens of their works may be found in most of the museums of South Germany; these almost self-taught artists display great dexterity in the management of their strange canvas, and considerable merit in the delicate manipulation of their pigments; sometimes they even imitate fine line engravings in pen

and ink without injuring the fragile surface. They
delight specially in treating subjects of traditional
interest, as Kaiser Max on the Martinswand, the
beautiful Philippine Welser, the heroic Hofer, and the
patron saints and particular devotions of their village
sanctuaries. Kranach's Mariähilf is thus an object of
most affectionate care. The 'web' is certainly like that
of no ordinary spider; but it is reported that this
family has cultivated a particular species for the purpose,
and an artist friend who had been in Mexico mentioned
to me having seen there spiders'-webs almost as solid
as these. I was not able, however, to learn any tradition of the importation of these spiders from Mexico.
In the first room on the second floor are to be seen the
characteristic letter written, as I have said, by Hofer,
shortly before his end, and other relics of him and the
other patriots, such as the hat and breviary of the
Franciscan Haspinger. Also an Italian gun taken by
the *Akademische Legion*—the band of loyal volunteer
students of Innsbruck university, in the campaign of
1848—and I think some trophies also of the success of
Tirolese arms against the attempted invasion of the
later Italian war, in which as usual the skill of these
people as marksmen stood them in good stead. Anyone
who wishes to judge of their practice may have plenty
of opportunity in Innsbruck, for their rifles seem to be
constantly firing away at the *Schiess-stand*; so constantly as to form an annoyance to those who are not
interested in the subject.

This *Schiess-stand,* or rifle-butt, was set up in 1863, in commemoration of the fifth centenary of Tirol's union with Austria and its undeviating loyalty. No history presents an instance of a loyalty more intimately connected with religious principle than the loyalty of Tirol; the two traditions are so inseparably interwoven that the one cannot be wounded without necessarily injuring the other. The present Emperor and Empress of Austria are not wanting to the devout example of their predecessors, but the modern theory of government leaves them little influence in the administration of their dominions. Meantime the anti-Catholic policy of the Central Government creates great dissatisfaction and uneasiness in Tirol. Other divisions of the empire had been prepared for such by laxity of manners and indifferentism to religious belief—the detritus, which the flood of the French revolution scattered more or less thickly over the whole face of Europe. But the valleys of Tirol had closed their passes to the inroads of this flood, and laws not having religion for their basis are there just as obnoxious in the nineteenth as they would have been in any former century.

In concluding my notice of the capital of Tirol, it may be worth while to mention that the census of January 1870 gives it a population (exclusive of military) of 16,810, being an increase of 2,570 over the twelve preceding years.

CHAPTER X.

NORTH TIROL—OBERINNTHAL.

INNSBRUCK TO ZIRL AND SCHARNITZ—INNSBRUCK TO THE LISENS-FERNER.

I taught the heart of the boy to revel
In tales of old greatness that never tire.
AUBREY DE VERE.

THOSE who wish to visit the legend-homes of Tirol without any great measure of 'roughing,' will doubtless find Innsbruck the most convenient base of operations for many excursions of various lengths to places which the pedestrian would take on his onward routes. Those on the north and east, which have been already suggested from Hall and Schwatz, may also be treated thus. It remains to mention those to be found on the west, north-west, and south. But first there is Mühlau, also to the east, reached by an avenue of poplars between the right bank of the Inn and the railway; where the river is crossed by a suspension-bridge. There are baths here which are much visited by the Innsbruckers, and many prefer staying there to Innsbruck itself. A pretty little new Gothic church adorns the height; the altar is bright with marbles of the country, and has a very creditable altar-piece by a Tirolean artist. Mühlau was celebrated in the *Befreiungskämpfe* through the courage of Baroness Sternbach, its chief resident; everywhere the patriots gathered she might have been found in their midst, fully armed and on her bold charger, inspiring all with courage. Arrested in her château at Mühlau during the Bavarian occupation, no threats or insult could wring from her any admission prejudicial to the interests of her country, or compromising to her son. She was sent to Munich, and kept a close prisoner there, as also were Graf Sarnthein and Baron Schneeburg, till the Peace of Vienna.

From either Mühlau or Innsbruck may be made the excursion to Frau Hütt, a curious natural formation which by a freak of nature presents somewhat the appearance of a gigantic petrifaction of a woman with a child in her arms. Of it one of the most celebrated of Tirolean traditions is told. In the time of Noe, says the legend, there was a queen of the giants living in these mountains, and her name was Frau Hütt. Nork makes out a seemingly rather far-fetched derivation for it out of the wife *der Behütete* (*i.e.* the behatted, or covered one), otherwise Odin, with the sky for his head-covering. However that may be, the legend says Frau Hütt had a son, a young giant, who wanted to cut down a pine tree to make a stalking-horse, but as the pine grew on the borders of a morass, he fell with his burden into the swamp. Covered over head and ears with mud, he came home crying to his mother, who ordered the nurse to wipe off the mud with fine crumb of white bread. This filled up the measure of Frau Hütt's life-long extravagance. As the servant approached, to put the holy gift of God to this profane use, a fearful storm came on, and the light of heaven was veiled by angry clouds; the earth rocked with fear, then opened a yawning mouth, and swallowed up the splendid marble palace of Frau Hütt, and the rich gardens surrounding it. When the sky became again serene, of all the former verdant beauty nothing remained; all was wild and barren as at present. Frau Hütt, who had run for refuge with her son in her arms to a neighbouring eminence, was turned into a rock. In place of our 'Wilful waste makes woeful want,' children in the neighbourhood are warned from waste by the saying, 'Spart eure Brosamen für die Armen, damit es euch nicht ergehe wie der Frau Hütt.'[1] Frau Hütt also serves as the popular barometer of Innsbruck; and when the old giantess appears with her 'night-cap' on, no one undertakes a journey. This excursion will take four or five hours. On the way, Büchsenhausen is passed, where, as I have already mentioned, Gregory Löffler cast the statues of the Hofkirche. I have also given already the legend of the Bienerweible. As a consequence of the state execution which occasioned her melancholy aberrations, the castle was forfeited to the crown. Ferdinand Karl, however, restored it to the family. It was subsequently sold, and became one

---

[1] Spare your bread for the poor, and escape the fate of Frau Hütt. See some legends forming a curious link between this, and that of Ottilia Milser in Stöber *Sagen des Elsasses*, pp. 257-8.

of the most esteemed breweries of the country, the cellars being hewn in the living rock; and its 'Biergarten' is much frequented by holiday-makers. Remains of the old castle are still kept up; among them the chapel, in which are some paintings worth attention. On one of the walls is a portrait of the Chancellor's son, who died in the Franciscan Order in Innsbruck, in his ninety-first year.

If time allows, the Weierburg and the Maria-Brünn may be taken in the way home, as it makes but a slight digression; or it may be ascended from Mühlau. The so-called Mühlauer Klamm is a picturesque gorge, and the torrent running through it forms some cascades. Weierburg affords a most delightful view of the picturesque capital, and the surrounding heights and valleys mapped out around. Schloss Weierburg was once the gay summer residence of the Emperor Maximilian, and some relics of him are still preserved there.

Hottingen, which might be either taken on the way when visiting Frau Hütt or the Weierburg, is a sheltered spot, and one of the few in the Innthal where the vine flourishes. It is reached by continuing the road past the little Church of Mariähilf across the Inn; it had considerable importance in mediæval times, and has consequently some interesting remains, which, as well as the bathing establishment, make it a rival to Mühlau. In the church (dedicated to St. Nicholas) is Gregory Löffler's monument, erected to him by his two sons. The Count of Trautmannsdorf and other noble families of Tirol have monuments in the Friedhof. The tower of the church is said to be a remnant of a Roman temple to Diana. To the right of the church is Schloss Lichtenthurm, well kept up, and often inhabited by the Schneeburg family. On the woody heights to the north is a little pilgrimage chapel difficult of access, and called the *Höttingerbilde*. It is built over an image of our Lady found on the spot in 1764, by a student of Innsbruck who ascribed his rapid advance in the schools to his devotion to it. On the east side of the Höttinger stream are some remains of lateral mining shafts, which afford the opportunity of a curious and difficult, though not dangerous, exploration. There are some pretty stalactitic formations, but on a restricted scale.

There is enough of interest in a visit to Zirl to make it the object of a day's outing; but if time presses it may be reached hence, by pursuing the main street of this suburb, called, I know not why, *zum grossen Herr-Gott*, which continues in a path along an almost direct line of about seven miles through field and forest, and for the last four or five following the bank of the Inn. Or the whole route may be taken in a carriage from Innsbruck, driving past the rifle-butt under Mariähilf. At a distance of two miles you pass Kranebitten, or Kranewitten, not far from which, at a little distance on the right of the road, is a remarkable ravine in the heights, which approach nearer and nearer the bank of the river. It is well worth while to turn aside and visit this ravine, which goes by the name of the *Schwefelloch*. It is an accessible introduction on a small scale to the wild and fearful natural solitudes we read of with interest in more distant regions. The uneven path is closed in by steep and rugged mountain sides, which spontaneously recall many a poet's description of a visit to the nether world. At some distance down the gorge, a flight of eight or nine rough and precarious steps cut in the rock, and then one or two still more precarious ladders, lead to the so-called *Hundskirche*, or *Hundskapelle*,[1] which is said to derive its name from having been the last resort of Pagan mysteries when heathendom was retreating before the advance of Christianity in Tirol. Further on, the rocks bear the name of the Wagnerwand (*Wand* being a wall), and the great and lesser Lehner; and here they seem almost to meet high above you and throw a strange gloom over your path, and the torrent of the *Sulz* roars away below in the distance; while the oft-repeated answering of the echo you evoke is more weird than utter silence. The path which has hitherto been going north now trends round to the west, and displays the back of the Martinswand, and the fertile so-called *Zirlerchristen*, soon affording a pleasing view both ways towards Zirl and Innsbruck. There is rough accommodation here for the night for those who would ascend the Gross Solstein, 9,393 feet; the Brandjoch, 7,628 feet; or the Klein Solstein, 8,018 feet—peaks of the range which keep Bavaria out of Tirol.

[1] The dog's church or chapel.

As we proceed again on the road to Zirl, the level space between the mountains and the river continues to grow narrower and narrower, but what there is, is every inch cultivated; and soon we pass the *Markstein* which constitutes the boundary between Ober and Unter-Innthal. By-and-by the mountain slopes drive the road almost down to the bank, and straight above you rises the foremost spur of the Solstein, the Martinswand, so called by reason of its perpendicularity, celebrated far and wide in *Sage* and ballad for the hunting exploit and marvellous preservation of Kaiser Max.

It was Easter Monday, 1490; Kaiser Max was staying at Weierburg, and started in the early morning on a hunting expedition on the Zirlergebrge. So far there is nothing very remarkable, for his ardent disposition and love of danger often carried him on beyond all his suite; but then came a marvellous accident, the accounts of the origin of which are various. There is no one in Innsbruck but has a version of his own to tell you. As most often reported, the chamois he was following led him suddenly down the very precipice I have described. The steepness of the terrible descent did not affright him; but in his frantic course one by one the iron spikes had been wrenched from his soles, till at last just as he reached a ledge, scarcely a span in breadth, he found he had but one left. To proceed was impossible, but—so also was retreat. There he hung, then, a speck between earth and sky, or as

Collin's splendid popular ballad, which I cannot forbear quoting, has it:—

> Hier half kein Sprung,
> Kein Adler-Schwung
> Denn unter ihm senkt sich die Martinswand
> Der steilste Fels im ganzen Land.
>
> Er starrt hinab
> In's Wolkengrab
> Und starrt hinaus in's Wolkenmeer
> Und schaut zurück, und schaut umher.
>
> *   *   *   *   *
>
> Wo das Donnergebrüll zu Füssen ihm grollt
> Wo das Menschengewühl tief unter ihm rollt:
> Da steht des Kaisers Majestät
> Doch nicht zur Wonne hoch erhöht.
>
> Ein Jammersohn
> Auf luft'gem Thron
> Findet sich Max nun plötzlich allein
> Und fühlt sich schaudernd, verlassen und klein.[1]

But the singers of the high deeds of Kaiser Max could not bring themselves to believe that so signal a danger could have befallen their hero by mere accident. They must discover for it an origin to connect it with his political importance. Accordingly they have said that the minions of Sigismund *der Münzreiche*, dispossessed

---

[1] His well-known daring, emulating that of the chamois and the eagle, was of no avail now; for straight under him sinks the Martin's Wall, the steepest cliff of the whole country-side.

He gazes down through that grave of clouds. He gazes abroad over that cloud-ocean. He glances around, and his gaze recoils.

With only the thunder-roll of the people's voices beneath, there stands the Kaiser's Majesty. But not raised aloft to receive his people's homage. A son of sorrow, on a throne of air, the great Maximilian all at once finds himself isolated, horror-stricken, and small.

at his abdication, had plotted to lead Max, the strong redresser of wrongs, the last flower of chivalry, the hope of the Hapsburg House, the mainstay of his century, into destruction; that it was not that the innocent chamois led the Kaiser astray, but that the conspirators misled him as to the direction it had taken.

Certainly, when one thinks of the situation of the empire at that moment, and of Hungary, the borderland against the Turks, suddenly deprived of its great King Matthias Corvinus, even while yet at war with them, only four days before [1]; when we think that the writers of the ballad had before their eyes the great amount of good Maximilian really did effect not only for Tirol, but for the empire and for Europe, and then contemplated the idea of his career being cut short thus almost at the outset, we can understand that they deemed it more consonant with the circumstances to believe so great a peril was incurred as a consequence of his devotion to duty rather than in the pursuit of pleasure.

Here, then, he hung; a less fearless hunter might have been overawed by the prospect or exhausted by the strain. Not so Kaiser Max. He not only held on steadfastly by the hour, but was able to look round him so calmly that he at last discerned behind him a cleft in the rock, or little cave, affording a footing less precarious than that on which he rested. The ballad may

---

[1] 'With him,' says a Hungarian ballad, 'Righteousness went down into the grave: and the Sun of Pest-Ofen sank towards its setting.'

be thought to say that it opened itself to receive him. The rest of the hunting party, even those who had nerve to follow him to the edge of the crag, could not see what had become of him. Below, there was no one to think of looking up; and if there had been, even an emperor could hardly have been discerned at a height of something like a thousand feet. The horns of the huntsmen, and the messengers sent in every direction to ask counsel of the most experienced climbers, within a few hours crowded the banks on both sides with the loyal and enthusiastic people; till at last the wail of his faithful subjects, which could be heard a mile off, sent comfort into the heart of the Kaiser, who stood silent and stedfast, relying on God and his people. Meantime, the sun had reached the meridian; the burning rays poured down on the captive, and gradually as the hours went by the rocks around him grew glowing hot like an oven. Exhausted by the long fast, no less than the anxiety of his position, and the sharp run that had preceded the accident, he began to feel his strength ebbing away. One desire stirred him—to know whether any help was possible before the insensibility, which he felt must supervene, overcame him. Then he bethought him of writing on a strip of parchment he had about him, to describe his situation, and to ask if there was any means of rescue. He tied the scroll to a stone with the cord of his hunting-horn, and threw it down into the depth. But no sound came in answer.

In the meantime all were straining to find a way of

escape. Even the old Archduke Sigismund who, though he is never accused of any knowledge of the alleged plot of his courtiers, yet may well be supposed to have entertained no very good feeling towards Maximilian, now forgot all ill-will, and despatched swift messengers to Schwatz to summon the cleverest *Knappen* to come with their gear and see if they could not devise a means for reaching him with a rope; others ran from village to village, calling on all for aid and counsel. Some rang the storm-bells, and some lighted alarm fires; while many more poured into the churches and sanctuaries to pray for help from on High; and pious brotherhoods, thousands in number, marching with their holy emblems veiled in mourning, and singing dirges as they came, gathered round the base of the Martinswand.

The Kaiser from his giddy height could make out something of what was going on : but as no answer came, a second and a third time he wrote, asking the same words. And when still no answer came—I am following Collin's imaginative ballad—his heart sank down within him and he said, 'If there were any hope, most surely my people would have sent a shout up to me. So there is no doubt but that I must die here.' Then he turned his heart to God, and tried to forget everything of this earth, and think only of that which is eternal. But now the sun sank low towards the horizon. While light yet remained, once more he took his tablet and wrote; he had no cord left to attach it to the stone, so he bound it with his gold chain—of what use were earthly ornaments any more to him?—'and threw it down,' as the ballad forcibly says, 'into the living world, out of that grave high placed in air.'

One in the crowd caught it, and the people wept aloud as he read out to them what the Kaiser had traced with failing hand. He thanked Tirol for its loyal interest in his fate; he acknowledged humbly that his suffering was a penance sent him worthily by heaven for the pride and haughtiness with which he had pursued the chase, thinking nothing too difficult for him. Now he was brought low. He offered his blood and his life in satisfaction. He saw there was no help to be hoped for his body; he trusted his soul to the mercy of God. But he besought them to send to Zirl, and beg the priest there to bring the Most Holy Sacrament and bless his last hour with Its Presence. When It arrived they were to announce it to him by firing off a gun, and another while the Benediction was imparted. Then he bid them all pray for steadfastness for him, while the pangs of hunger gnawed away his life.

The priest of Zirl hastened to obey the summons, and the Kaiser's injunctions were punctually obeyed. Meantime, the miners of Schwatz were busy arranging their plan of operations—no easy matter, for they stood fifteen hundred feet above the Emperor's ledge. But before they were ready for the forlorn attempt, another deliverer appeared upon the scene with a strong arm, supported the almost lifeless form of the Emperor—for he had now been fifty-two hours in this sad plight—and bore him triumphantly up the pathless height. There he restored him to the people, who, frantic with joy, let him pass through their midst without observing his appearance. Who was this deliverer? The traditions of the time say he was an angel, sent in answer to the Kaiser's penitential trust in God and the prayers of the people. Later narrators say—some, that he was a bold huntsman; others, a reckless outlaw to whom the track was known, and these tell you there is a record of a pension being paid annually in reward for the service, if not to him, at least to some one who claimed to have rendered it.[1]

The Monstrance, which bore the Blessed Sacrament from Zirl to carry comfort to the Emperor in his dire need, was laid up among the treasures of Ambras.

Maximilian, in thanksgiving for his deliverance, resolved to be less reckless in his future expeditions, and never failed to remember the anniversary. He also employed miners from Schwatz to cut a path down to the hole, afterwards called the Max-Höhle, which had sheltered him, to spare risk to his faithful subjects, who *would* make the perilous descent to return thanks on the spot for his recovery; and he set up there a crucifix,

---

[1] Primisser, who took great pains to collect all the various traditions of this event, mentions a favourite huntsman of the Emperor, named Oswald Zips, whom he ennobled as Hallaurer v. Hohenfelsen. This may have been the actual deliverer, or may have been supposed to be such, from the circumstance of the title being Hohenfelsen, or Highcliff; and that a patent of nobility was bestowed on a huntsman would imply that he had rendered *some* singular service: the family, however, soon died out.

with figures of the Blessed Virgin and S. John on either side large enough to be seen from below; and even to the present day men used to dangerous climbing visit it with similar sentiments. It is not often the tourist is tempted to make the attempt, and they must be cool-headed who would venture it. The best view of it is to be got from the remains of the little hunting-seat and church which Maximilian afterwards built on the Martinsbühl, a green height opposite it, and itself no light ascent. It is said Maximilian sometimes shot the chamois out of the windows of this villa. The stories are endless of his hardihood and presence of mind in his alpine expeditions. At one time, threatened by the descent of a falling rock, he not only was alert enough to spring out of the way in time, but also seized a huntsman following him, who was not so fortunate, and saved him from being carried over the precipice. At another he saw a branch of a tree overhanging a yawning abyss; to try his presence of mind he swung himself on to it, and hung over the precipice; but crack! went the branch, and yet he saved himself by an agile spring on to another tree. Another time, when threatened by a falling rock, his presence of mind showed itself in remaining quite still close against the mountain wall, in the very line of its course, having measured with his eye that there was space enough for it to clear him. But enough for the present.

Zirl affords a good inn and a timely resting-place, either before returning to Innsbruck, or starting afresh to visit the Isarthal and Scharnitz. The ascent of the Gross Solstein is made from Zirl, as may also be that of the Martinswand. In itself Zirl has not much to arrest attention, except its picturesque situation (particularly that of its 'Calvarienberg,' to form which the living rocks are adapted), and its history, connecting it with the defence of the country against various attacks from Bavaria. Proceeding northwards along the road to Seefeld, and a little off it, you come upon Fragenstein, another of Maximilian's hunting-seats, a strong fortress for some two hundred years before his time, and now a fine ruin. There are many strange tales of a great treasure buried here, and a green-clad huntsman, who appears from time to time, and challenges the peasants to come and help him dig it out, but something always occurs to prevent the successful issue of the adventure. Once a party of excavators got so far that they saw the metal vessel enclosing it; but then suddenly arose such a frightful storm, that none durst proceed with the work; and after that the clue to its place of concealment was lost. Continuing the somewhat steep ascent, Leiten is passed, and then Reit, with nothing to arrest notice; and then Seefeld, celebrated by the legend my old friend told me on the Freundsberg.[1] The Archduke Ferdinand built a special chapel to the left of the parish church, called *die Heilige Blutskapelle*, in 1575, to contain the Host which had convicted Oswald Milser, and which is even

---

[1] See chapter on Schwatz.

now an object of frequent pilgrimage. The altar-piece was restored last year very faithfully, and with conderable artistic feeling, by Haselwandter, of Botzen. It is adorned with statues of the favourite heroes of the Tirolese legendary world, St. Sigismund and St. Oswald, and compartment bas-reliefs of subjects of Gospel history known as 'the Mysteries of the Rosary.' The tone of the old work has been so well caught, that it requires some close inspection to distinguish the original remains from the new additions. The Archduchess Eleonora provided the crystal reliquary and crown, and the rich curtains within which it is preserved. At a little distance to south-west of Seefeld, on a mountain-path leading to Telfs, is a little circular chapel, built by Leopold V. in 1628, over a crucifix which had long been honoured there. It is sometimes called the Kreuz-kapelle, but more often the *zur-See-kapelle*, though one of the two little lakes, whence the appellation, and the name of Seefeld too, was derived, dried out in 1807. There is also a legend of the site having been originally pointed out by a flight of birds similar to that I have given concerning S. Georgenberg.

The road then falls more gently than on the Zirl side, but is rugged and wild in its surroundings, to Scharnitz, near which you meet the blue-green gushing waters of the Isar. Scharnitz has borne the brunt of many a terrible contest in the character of outpost of Tirolean defences; it is known to have been a fortress in the time of the Romans. It was one of the points strengthened by Klaudia de' Medici, who built the 'Porta Klaudia' to command the pass. Good service it did on more than one occasion; but it succumbed in the inroad of French and Bavarians combined, in 1805. It was garrisoned at that time by a small company of regular troops, under an English officer in the Austrian service named Swinburne, whose gallant resistance was cordially celebrated by the people. He was overwhelmed, however, by superior numbers and appliances, and at Marshal Ney's orders the fort was so completely destroyed, that scarcely a trace of it is now to be found.[1]

---

[1] *To the Editor of the* 'Monthly Packet.'

Sir,—I think it possible that R. H. B. (to whom we owe the very interesting *Traditions of Tirol*), and perhaps others of your readers, may care to hear some of the particulars, as they are treasured by his family, of the defence of Scharnitz by Baron Swinburne. R. H. B. speaks of it in your number of last month. That defence was so gallant as to call forth the respect and admiration even of his enemies, and Baron Swinburne was given permission to name his own terms of surrender.

He requested for himself, and those under him, that they might be allowed to retain their swords. This was granted, and the prisoners were sent to Aix-la-Chapelle, where everyone was asking in astonishment who were 'les prisonniers avec l'épée a côté.'

The Eagles of Austria, that had been so nobly defended by the Englishman and his little band, never fell into the hands of the French. One of the Tirolese escaped, with the colours wrapped round his body under his clothes, and though he was hunted among the mountains for months, he was never taken; and some years after he came to his commander in Vienna and gave him the colours he had so bravely defended. They are now in possession of Baron Edward Swinburne, the son of the defender of Scharnitz, who himself won, before he was eighteen, the Order of 'the Iron Crown,' by an act that well deserves to be called 'a golden deed;' and ere he was twenty he had led his first

It is the border town against Bavaria, and is consequently enlivened by a customs office and a few uniforms, but it is a poor place. I was surprised to be accosted and asked for alms by a decent-looking woman, whom I had seen kneeling in the church shortly before: as this sort of thing is not common in Tirol. She told me the place had suffered sadly by the railway; for before, it was the post-station for all the traffic between Munich and Innsbruck and Italy. The industries of the place were not many or lucrative; the surrounding forests supply some employment to woodmen; and what she called *Dirstenöhl*, which seems to be dialectic for *Steinhöl* or petroleum, is obtained from the bituminous soil in the neighbourhood; it is obtained by a kind of distillation—a laborious process. The work lasted from S. Vitus' Day to the Nativity of the Blessed Virgin; that was now past, and her husband, who was employed in it, had nothing to do: she had an old father to support, and a sick child. Then she went on to speak of the devotion she had just been reciting in

and last forlorn hope, when he received so severe a wound as to cost him his leg, which has incapacitated him for further service.

His father received the highest military decoration of Austria, that of 'Maria Teresa;' he fought at Austerlitz and Wagram; on the latter occasion he was severely wounded. Later in life, he was for many years Governor of Milan.

Hoping that a short record of true and faithful services performed by Englishmen for their adopted country, may prove of some interest to your readers, and with many thanks to R. H. B. for what has been of so much interest to us,

I am, Sir, yours faithfully,

September, 1870.   A. SWINBURNE.

the church to obtain help, and evidently looked upon her meeting with me as an answer to it. It seemed to consist in saying three times, a petition which I wrote down at her dictation as follows:—' Gott grüsse dich Maria! ich grüsse dich drei und dreizig Tausand Mal; O Maria ich grüsse dich wie der Erzengel Gabriel dich gregüsset hat. Es erfreuet dich in deinen Herzen dass der Erzengel Gabriel den himmlischen Gruss zu dir gebracht hat. Ave Maria, &c.' She said she had never used that devotion and failed to obtain her request. I learnt that the origin she ascribed to it was this:—A poor girl, a cow-herd of Dorf, some miles over the Bavarian frontier, who was very devout to the Blessed Virgin, had been in the habit while tending her herds of saying the rosary three times every day in a little Madonna chapel near her grazing-ground. But one summer there came a great heat, which burnt up all the grass, and the cattle wandered hither and thither seeking their scanty food, so that it was all she could do to run after and keep watch over them. The good girl was now much distressed in mind; for the tenour of her life had been so even before, that when she made her vow to say the three rosaries, it had never occurred to her such a contingency might happen. But she knew also that neither must she neglect her supervision of the cattle committed to her charge. While praying then to Heaven for light to direct her in this difficulty, the simple girl thought she saw a vision of our Lady, bidding her be of good heart, and she would teach her a prayer to say instead, which would not take as long as the rosary, and would please her as well, and that she should teach it also to others who might be overwhelmed with work like herself. This was the petition I have quoted above. But the maid was too humble to speak of having received so great a favour, and lived and died without saying anything about it. When she came to die, however, her soul could find no rest, for her commission was unfulfilled; and whenever anyone passed alone by the wayside chapel where she had been wont to pray, he was sure to see her kneeling there. At last a pious neighbour, who knew how good she had been, summoned courage to ask her how it was that she was dealt with thus. Then the good girl told him what had befallen her long ago on that spot, and bid him fulfil the part she had neglected, adding, 'But tell them also not to think the mere saying the words is enough; they must pray with faith and dependence on God, and also strive to keep themselves from sin.'

In returning from Zirl to Innsbruck, the left bank may be visited by taking the Zirl bridge and pursuing the road bordering the river; you come thus to Unterperfuss, another bourne of frequent excursion from Innsbruck, the inn there having the reputation of possessing a good cellar, and the views over the neighbourhood being most romantic, the *Château* of Ferklehen giving interest to the natural beauties around. Hence, instead of pursuing the return journey at once, a digression may be made through the Selrainthal (Selrain, in the dialect of the neighbourhood, means the edge of a mountain); and it is indeed but a narrow strip bordering the stream—the Melach or Malk, so called from its milk-white waters—which pours itself out by three mouths into the Inn at the debouchure of the valley. There is many a 'cluster of houses,' as German expresses [1] a settlement too small to be dignified with the name of village, perched on the heights around, but all reached by somewhat rugged paths. The first and prettiest is Selrain, which is always locally called Rothenbrunn, because the iron in the waters, which form an attraction to valetudinarian visitors, has covered the soil over which they flow with a red deposit. Small as it is, it boasts two churches, that to S. Quirinus being one of the most ancient in Tirol. The mountain path through the Fatscherthal, though much sought by Innsbruckers, is too rough travelling for the ordinary tourist, but affords a fine mountain view, including the magnificent *Fernerwand*, or glacier-wall, which closes it in, and the three shining and beautifully graduated peaks of the Hohe Villerspitz. At a short distance from Selrain may be found a pretty cascade, one of the six falls of the Saigesbach. Some four or five miles further along the valley is one of the numerous villages named Gries; and about five miles more of

[1] *Häusergruppe.*

mountain footpath leads to the coquettishly perched sanctuary of St. Sigismund, the highest inhabited point of the Selrainthal. It is one of the many high-peaked buildings with which the Archduke Sigismund, who seems to have had a wonderful eye for the picturesque, loved to set off the heaven-pointing cones of the Tirolese mountains. Another opening in the mountains, which runs out from Gries, is the Lisenthal, in the midst of which lie Juvenan and Neurätz, the latter much visited by parties going to pick up the pretty crystal spar called 'Andalusiten.' Further along the path stands by the wayside a striking fountain, set up for the refreshment of the weary, called the Magdalenenbründl, because adorned with a statue of the Magdalen, the image of whose penitence was thought appropriate to this stern solitude by the pious founder. The Melach is shortly after crossed by a rustic bridge, and a path over wooded hills leads to the ancient village of Pragmar. Hence the ascent of the Sonnenberg or Lisens-Ferner is made. The monastery of Wilten has a summer villa on its lower slope, serving as a dairy for the produce of their pastures in the neighbourhood; a hospitable place of refreshment for the traveller and alpine climber, and with its chapel constituting a grateful object both to the pilgrim and the artist. The less robust and enterprising will find an easier excursion in the Lengenthal, a romantically wild valley, which forms a communication between the Lisenthal and the Œtzthal.

The Selrainthalers are behind none in maintaining the national character. When the law of conscription—one of the most obnoxious results of the brief cession to Bavaria—was propounded, the youths of the Selrain were the first to show that, though ever ready to devote their lives to the defence of the fatherland, they would never be enrolled in an army in whose ranks they might be sent to fight in they knew not what cause—perhaps against their own brethren. The generous stand they made against the measure constituted their valley the rendezvous of all who would escape from it for miles round, and soon their band numbered some five hundred. During the whole of the Bavarian occupation they maintained their independence, and were among the first to raise the standard of the year 1809. A strong force was sent out on March 14 to reduce them to obedience, when the Selrainers gave good proof that it was not cowardice which had made them refuse to join the army. They repulsed the Bavarian regulars with such signal success, that the men of the neighbourhood were proud to range themselves under their banner, which as long as the campaign lasted was always found in the thickest of the fight. No less than eleven of their number received decorations for personal bravery. In peace, too, they have shown they know how to value the independence for which they fought; though their labours in the field are so greatly enhanced by the steepness of the ground which is their portion, that the men yoke themselves to the plough, and carry burdens over places where no oxen could be guided. Their industry and perseverance provide them so well with enough to make them contented, if not prosperous, that ' *in Selrain hat jeder zu arbeiten und zu essen*' (in Selrain there is work and meat enough for all) is a common proverb. The women, who are unable for the reason above noted to take so much part in field-labours as in some other parts, have found an industry for themselves in bleaching linen, and enliven the landscape by the cheerful zest with which they ply their thrifty toil.

The path for the return journey from Selrain to Ober-Perfuss—or foot of the upper height—is as rugged as the other paths we have been traversing, but is even more picturesque. The church is newly restored, and contains a monument, with high-sounding Latin epitaph, to one Peter Anich, of whose labours in overcoming the difficulties of the survey and mensuration of his country, which has nowhere three square miles of plain, his co-villagers are justly proud. He was an entirely self-taught man, but most accurate in his observations, and he induced other peasants to emulate his studies. Ober-perfuss also has a mineral spring. A pleasant path over hills and fields leads in about an hour to Kematen, a very similar village; but the remains of the ruined hunting-seat of Pirschenheim, now used as an ordinary lodging-house, adds to its picturesqueness. Near by it may also be visited the pretty waterfall of the Sendersbach. A shorter and easier stage is the next, through the fields to Völs or Vels, which clusters at the foot of the Blasienberg, once the dwelling of a hermit, and still a place of pilgrimage and the residence of the priest of the village. The parish church of Vels is dedicated in honour of S. Jodok, the English saint, whose statue we saw keeping watch over Maximilian's tomb at Innsbruck. Another hour across the level ground of the Galwiese, luxuriantly covered with Indian corn, brings us back to Innsbruck through the Innrain; the Galwiese has its name from the echo of the hills, which close in the plain as it nears the capital; *wiese* being a meadow, and *gal* the same form of *Schall*—resonance, which occurs in *Nachtigall*, nightingale; and also, strangely enough, in *gellen*, to sound loudly (or yell). At the cross-road (to Axams) we passed some twenty minutes out of Völs, where the way is still wild, is the so-called *Schwarze Kreuz-kapelle*. One Blasius Hölzl, ranger of the neighbouring forest, was once overtaken by a terrible storm; the Geroldsbach, rushing down from the Götzneralp, had obliterated the path with its torrents; the reflection of each lightning flash in the waste of waters around seemed like a sword pointed at the breast of his horse, who shied and reared, and threatened to plunge his rider in the ungoverned flood. Hölzl was a bold forester, but he had never known a night like this; and as the rapidly succeeding flashes almost drove him to distraction, he vowed to record the deliverance on the spot by a cross of iron, of equal weight to himself and his mount, if he reached his fireside in safety. Then suddenly the noisy wind subsided, the clouds owned themselves spent, and in place of the angry forks of flame only soft and friendly sheets of light played over the country, and enabled him to steer his homeward way. Hölzl kept his promise, and a black metal cross of the full weight promised long marked the spot, and gave it its present name.[1] The accompanying figures of our Lady and S. John having subsequently been thrown down, it was removed to the chapel on Blasienberg. Ferneck, a pleasant though primitive bath establishment, is prettily situated on the Innsbruck side of the Galwiese, and the church there was also once a favourite sanctuary with the people; but when the neighbouring land was taken from the monks at Wilten, who had had it ever since the days of the penitent giant Haymon, it ceased to be remembered.

Starting from Innsbruck again in a southerly direction, a little beyond Wilten, already described, we

---

[1] Such offerings are met with in other parts of Tirol: in one place we shall find a candle offered of equal weight to an infant's body. They present a striking analogy with the Sanskrit *tulâdâna* or weight-gift; the practice of offering to a temple or Buddhist college a gift of silver or even gold of the weight of the offerer's body appears not to have been infrequent and tolerably ancient. Lassen (*Indische Alterthumskünde*, vol. iii. p. 810) mentions an instance of the *revival* of the custom by a king named Shrikandradeva, who offered the weight of his own body in gold to the temple at Benares (*circa* 1025); and (vol. iv. p. 373) another in which Aloungtsethu, King of Birmah, in 1101, made a similar offering in silver to a temple which he built at Buddhagayâ. He refers also to earlier instances 'in H. Burney's note 19 in *As. Res.* vol. xx. p. 177, and one by Fell in *As. Res.* vol. xv. p. 474.'

reach Berg Isel. Though invaded in part by the railway, it is still a worthy bourne of pilgrimage, by reason of the heroic victories of the patriots under Hofer. On Sunday and holiday afternoons parties of Innsbruckers may always be found refreshing these memories of their traditional prowess. It is also precious on less frefrequented occasions for the splended view it affords of the whole Innthal. Two columns in the *Scheisstand* record the honours of April 29 and August 30, 1809, with the inscription, '*Donec erunt montes et saxa et pectora nostra Austriacae domini moenia semper erunt.*' I must confess, however, that the noise of the perpetual rifle-practice is a great vexation, and prevents one from preserving an unruffled memory of the patriotism of which it is the exponent; but this holds good all over Germany. Here, on May 29, fell Graf Johan v. Stachelburg, the last of his noble family, a martyr to his country's cause. The peasants among whom he was fighting begged him not to expose his life so recklessly, but he would not listen. 'I shall die but once,' he replied to all their warnings; 'and where could it befall me better than when fighting for the cause of God and Austria?' He was mortally wounded, and carried in a litter improvised from the brushwood to Mutters, where he lies buried. A little beyond the southern incline of Berg Isel a path strikes out to the right, and ascends the heights to the two villages of Natters and Mutters, the people of which were only in 1786 released from the obligation of going to Wilten for their Mass of obligation.

Natters has some remains of one of Archduke Sigismund's high-perched hunting-seats, named Waidburg; he also instituted in 1446 a foundation for saying five Masses weekly in its chapel.

There are further several picturesque mountain walks to be found in the neighbourhood of Innsbruck, under the grandly towering Nockspitze and the Patscherkofl. Or again from either Mutters or Natters there is a path leading down to Götzens, Birgitz, Axams, and Grintzens, across westwards to the southern end of the Selrainthal. Götzens (from *Götze*, an idol), like the Hundskapelle, received its name for having retained its heathen worship longer than the rest of the district around. The ruins, which you see on a detached peak as you leave Götzens again, are the two towers of Liebenberger, and Völlenberger the poor remains of Schloss Völlenberg, the seat of an ancient Tirolean family of that name, who were very powerful in the twelfth and thirteenth centuries. It fell in to the Crown during the reign of Friedrich *mit der leeren Tasche*, by the death of its last male heir. Frederick converted it into a state-prison. The noblest person it ever harboured was the poet Oswald von Wolkenstein. Himself a knight of noble lineage, he had been inclined in the early part of Frederick's reign to join his influence with the rest of the nobility against him, because he took alarm at his familarity with the common people. Frederick's sudden establishment of his power, and the energetic proceedings he immediately adopted for consolidating it, took many by surprise, Oswald von Wolkenstein among the rest. He was a bard of too sweet song, however, to be shut up in a cage, and Friedl was not the man to keep the minstrel in durance when it was safe to let him be at large. He had no sooner established himself firmly on the throne than he not only released the poet, but forgetting all cause of animosity against him, placed him at his court, and delighted his leisure hours with listening to his warbling. Oswald's wild and adventurous career had stored his mind with such subjects as Friedl would love to hear sung. But we shall have more to say of Oswald when we come to his home in the Grödnerthal.

The next village is Birgitz; and the next, after crossing the torrent which rushes down from the Alpe Lizum, is Axams, one of the most ancient in the neigbourhood, after passing the opening to the lonesome but richly pastured Sendersthal, the slopes of which meet those of the Selrainthal.

The only remaining valley of North Tirol which I have room here to treat is the Stubay Thal.[1] Of the three or four ways leading into it from Innsbruck, all rugged, the most remarkable is called by the people 'beim Papstl' because that traversed by Pius VI. when he passed through Tirol, as I have already narrated. The first place of any interest is Waldrast, a pilgrim's chapel, dating from the year 1465. A poor peasant was

---

[1] I have occasion to give one of the most remarkable legends of the Oetzthal in the chapter on Wälsch-Tirol.

directed by a voice he heard in his sleep to go to the woods (*Wald*), and lay him down to rest (*Rast*), and it would be told him what he should do; hence the name of the spot. There the Madonna appeared to him, and bid him build a chapel over an image of her which appeared there, no one knew how, some years before.[1] A Servite monastery, built in 1624 on the spot, is now in ruins, but the pilgrimage is still often made. It may be reached from the railway station of Matrey. The ascent of the Serlesspitz being generally undertaken from here, it is called in Innsbruck the Waldrasterspitz. Fulpmes is the largest village of the Stubay Thal. The inhabitants are all workers in iron and steel implements, and among other things are reckoned to make the best spikes for the shoes of the mountain climbers. Their works are carried all over Austria and Italy, but less now than formerly. In the church are some pictures by a peasant girl of this place. Few will be inclined to pursue this valley further; and the only remaining place of any mark is Neustift, the marshy ground round which provides the Innsbruck market with frogs. The church of Neustift was built, at considerable cost, in the tasteless style of the last century. The wood carvings by the Tirolean artists Keller, Hatter, and Zatter, however, are meritorious.

[1] See a somewhat similar version in Nork's *Mythologie der Volksagen*, pp. 895–7.

CHAPTER XI.

WÄLSCH-TIROL.

THE WÄLSCH-TIROLISCHE ETSCHTHAL AND ITS TRIBUTARY VALLEYS.

It is not some Peter or James who has written these stories for a little circle of flattering contemporaries; it is a whole nation that has framed them for all times to come, and stamped them with the impress of its own mighty character.—AKSHAROUNIOFF, *Use of Fairy Tales.*

IT is time that we turn our attention to the Traditions of South and Wälsch-Tirol, though it must not be supposed that we have by any means exhausted those of the North. There are so many indications that ere long the rule over the province, or *Kreis*,[1] as it is called, of Wälsch-Tirol, may some day be transferred to Italy, that, especially as our present view of it is somewhat retrospective, it is as well to consider it first, and before its homogeneity with the rest of the principality is destroyed.[2]

Wälsch, or Italian-Tirol sometimes, especially of late, denominated the Trentino, comprises the sunniest, and some at least, of the most beautiful valleys of Tirol. The Etschthal, or valley of the Adige, which takes its source from the little lake Reschen, also

---

[1] Circle.

called *der Grüne,* from the colour of its waters, near Nauders, traverses both South and Wälsch-Tirol. That part of the Etschthal belonging to the latter *Kreis* takes a direct north to south direction down its centre. There branch out from it two main lines of valleys on the west, and two on the east. The northernmost line on the west side is formed of the Val di Non and the Val di Sole; on the east, of the Avisio valley under its various changes of name which will be noted in their place. The Southern line on the west is called Giudicaria, and on the east, Val Sugana, or valley of the Euganieren.

The traveller's first acquaintance with the *Wälsch-tirolische-Etschthal* will probably, as in my own case, be made in the Val Lagarina, through which the railway of Upper Italy passes insensibly on to Tirolese soil, for you are allowed to get as far as Ala before the custom-house visitation reminds you that you have passed inside another government. It is a wild gorge along which you run, only less formidable than that which you saw so grimly close round you as you left Verona. If you could but lift that stony veil on your left, you would see the beautiful Garda-See sparkling beside you; but how vexatious soever the denial, the envious mountains interpose their stern steeps to conceal it. Their recesses conceal too, but to our less regret, the famous field of Rivoli.

Borghetto is the first village on Tirolese soil, and Ala, in the Middle Ages called Sala, the first town. It thrives on the production of silk, introduced here from Lombardy about 1530. It has a picturesque situation, and some buildings that claim a place in the sketch-book. The other places of interest in the neighbourhood are most conveniently visited from Roveredo, or Rofreit as the Germans call it, a less important and pleasing town than Trent, but placed in a prettier neighbourhood. It received its name of Roboretum from the Latins, on account of the immense forests of oak with which it was surrounded in their time. The road leading through it, being the highway into the country, bristles all along its way with ancient strong-holds, as Avio, Predajo, Lizzana, Castelbarco, Beseno, and others, which have all had their share in the numerous struggles for ascendancy, waged for so many years between the Emperor, the Republic of Venice, the Bishops of Trent, and the powerful families inhabiting them. The last-named preserves a tradition of more peaceful interest. At the time that Dante was banished from Florence, Lizzana was a seat of the Scaligers, and they had him for their visitor for some time during his wanderings. Not far from it is the so-called Slavini di San Marco, a vast *Steinmeer*, which seems, as it were, a ruined mountain, such vast blocks of rock lie scattered on every side. There is little doubt the poet has immortalized the scene he had the opportunity of contemplating here in his description of the descent to the Inferno, opening of Canto XII. It is said that a fine city, called San

Marco, lies buried under these gigantic fragments, concerning which the country people were very curious, and were continually excavating to arrive at the treasure it was supposed to contain, till one day a peasant thus engaged saw written in fiery letters on one vast boulder, 'Beati quelli che mi volteranno' (happy they who turn me round). The peasant thought his fortune was made. There could be no doubt the promised happiness must consist in the riches which, turning over the stone should disclose. Plenty of neighbours were ready to lend a hand to so promising a toil; and after the most unheard-of exertions, the monster stone was upheaved. But instead of a treasure they found nothing but another inscription, which said 'Bene mi facesti, perchè le coscie mi duolevano (you have done me a good turn, for I had a pain in my ribs).[1] As the peasants felt no great satisfaction in working with no better pay than this, the buried city of San Marco ceased from this time to be the object of their search. Nevertheless, near Mori, on the opposite (west) side of the river, is a deep cave called 'la Busa del Barbaz,' concerning which the saying runs, that it was, ages ago, the lurking-place of a cruel white-bearded old man, who lived on human flesh, and that

---

[1] The sunnier and less thoughtful tone of mind in which the Italian particularly differs from the German character, is often to be traced in their legendary stories. Those of the Germans are nearly always made to convey some moral lesson; this is as often wanting in those of the Italians, who seem satisfied with making them means of amusement, without caring that they should be a medium of instruction.

whoso has the courage to explore the cave and discover his remains, will, immediately on touching them, be confronted by his spirit, who will tell the adventurous wight where an immense treasure lies hid. Some sort of origin for this fable may be found in an older tradition, which tells that idols, whose rites demanded human sacrifices, were cast down this cave by the first Christian converts of the Lenothal. The Slavini are closed by a rocky gorge, characteristically named Serravalle; and as the country again opens out another cave on the east bank is pointed out, which was for long years a resort of robbers, who plundered all who passed that way. These were routed out by the Prince-bishop of Trent in 1197, and a hospice for the relief of travellers built on the very spot which so long had been the terror of the wayfarer. The chapel was dedicated in honour of S. Margaret, and still retains the name.

Roveredo itself is crowned by a fort—Schloss Junk, or Castel nuovo—which has stood many a siege, originally built by the Venetians; but it is more distinguished by its villas and manufactories. The silk trade was introduced here in 1580, and has continuously added to the prosperity of the place. Gaetano Tacchi established relations with England at the end of the last century, and the four brothers of the same name, who now represent her house, are the richest family in Roveredo. They have a very pretty family vault near the Madonna del Monte, a pilgrimage reached by a road which starts behind the *Pfarrkirche* of Sta. Maria.

Another pilgrimage church newly established is the Madonna de Saletto. While the silk factories occupy the Italian hands, the Germans resident in Roveredo find employment at a newly-established tobacco factory. Much tobacco is grown in the Trentino.

A great deal of activity is seen in Roveredo. The *Corso nuovo* is a broad handsome street with fine trees. A new and handsome road, between the town and railway station, was laid out in the autumn of 1869. Outside the town is the so-called Lenoschlucht, reached by the *Strada nuova*, which crosses it by a daring high arched bridge. The cliff rises sheer on the right hand, and overlooking the dangerous precipice is the little chapel of S. Columban, seemingly perched there by enchantment. It is built over the spot where a hermit, who was held in veneration by the neighbourhood, had his retreat.

There are seven churches, but not much to remark in any of them. That of S. Rocchus was built in consequence of a vow made by the townspeople during the plague of 1630, to invite a settlement of Franciscans if it was stayed. The altar-piece is ascribed to Giovanni da Udine. There are several educational establishments, and a club which is devoted to propagandism of Italian tendencies.

The time to see Trent to advantage is in the month of June, not only for the sake of the natural beauties of climate and scenery, but because then falls the festa of S. Vigilius (26th), the evangelizer of the country, and the churches are crowded with all the surrounding mountain population, who, after religious observances have been duly fulfilled, indulge in all their characteristic games and amusements, often in representations of sacred dramas,[1] and always wind up with their favourite and peculiar illumination of their mountain sides by disposing bonfires in devices over a whole slope. This custom is the more worth noting that it is thought to be a remnant of fire-worship, prevailing before the entrance of the Etruscans.[2]

That their city was the see of S. Vigilius, and the seat of the great council of the Church, are reckoned by its people their greatest glories; and they delight to trace a parallel between their city and 'great Rome.' They reckon that it was founded in the time of Tarquinius Priscus by a colony of Etruscans, under a leader

---

[1] The Passion Plays of the Brixenthal, however, are reckoned the best. The performers gather and rehearse in the spring, and go round from village to village through the summer months, only, as amphitheatres are improvised in the open.

[2] It may be worth mentioning, as an instance of how the contagion of popular customs is transmitted, that on enquiring into some very curious grotesque ceremonies performed in Trent at the close of the carneval, and called its 'burial,' I learnt that it did not appear to be a Tirolean custom, but had been introduced by the soldiers of the garrison who, for a long time past, had been taken from the Slave provinces of the Austrian Empire, and thus a Slave popular custom has been grafted on to Tirol. Wälsch-Tirol, however, has its own customs for closing the carneval, too. In some places it is burnt in effigy; in some, dismissed with the following dancing-song (Schnodahüpfl) greeting,

<blockquote>
Evviva carneval!
Chelige manca ancor el sal;
El carneval che vien
Lo salerem più ben!
</blockquote>

named Rhætius, who established there the worship of Neptune, whence the name of Tridentum or Trent. That they occupied and fortified the country, and subsequently became a power formidable to the Empire; But some twenty-five years before the Christian era, Rhætia, as the country round was called, was conquered by Drusus, son-in-law of Augustus, and colonized. An ancient inscription preserved in the Schloss Buon Consiglio shows that Trent was the centre of the local government, which was exactly modelled on that of Rome. S. Vigilius, who spread the light of the faith here, was a born Roman, and suffered martyrdom in a persecution emulating those of Rome in the year 400. The city endured sieges and over-running from many of the barbarous nations which over-ran and sacked Rome, and researches into the ancient foundations show that the accumulation of ruins has raised the soil, as in Rome, some feet above the original ground plan— Ranzi says more than four metres. The traces of three distinct lines of walls, showing just as in Rome the progressive enlargement of the city, have been found, as also remains of a considerable amphitheatre, and many of inlaid pavements, &c., showing that it was handsomely built and provided. To complete the parallel, it was under the régime of an ecclesiastical ruler that, after years of distress and turmoil, its peace and prosperity were restored. The Bishop of Trent still retains his title of Prince, but the deprivation of his territorial rule was one of the measures of secularization of Joseph II.

There are sixteen churches in Trent, of which the most considerable is the Cathedral, dating from the eleventh century—with some remnants of sculpture, as the Lombard ornaments of the three porches, reckoned to belong to the seventh or eighth—a Romanesque building of massive design, built of the reddish-brown marble which abounds in the neighbourhood, with a Piazza and fountain before it. The interior is extensively decorated with frescoes. It is dedicated to S. Vigilius, whose relics are preserved in a silver sarcophagus. Among its other notabilia are a Madonna, by Perugino, and some good paintings of less esteemed masters; also a copy of the Madonna di San Luca of the Pantheon, presented in 1465 to the then Bishop of Trent, while on a visit to Rome, by the Pope, and ever since an object of popular veneration. As a curiosity, is shown a waxen image of the Blessed Virgin, modelled by a Jew. It also contains several curious brass monuments. The Church of Sta. Maria Maggiore, where the great Council was held, on this account, surpasses it in interest, though of small architectural merit. There is a legend that when the final Te Deum at the close of the Council was sung on December 4, 1563,[1] a crucifix, still pointed out in one of the side chapels[2] of the Cathedral, was seen to bow

---

[1] A centenary celebration of the Council was held at Trent in 1863, at which the late lamented Cardinal von Reisach presided as legate *a latere*.

[2] This chapel has lately been restored by Loth of Munich.

its head as if in token of approval of the constitutions that had been established. Sta. Maria Maggiore contains a picture of the Council, with the fathers in full session, which is not without interest, as all the costumes can still be made out, though quaint and faded and injured by lightning. It has also a very fine organ, the tone of which was so much esteemed at the time it was built, that it is said the Town Council determined to put out the eyes of the organ-builder,[1] lest he should endow any other city with as perfect an instrument. The meister, finding he could not prevail on the councilmen to relent, asked as a last favour to be allowed to play on his organ, which was willingly conceded; but as soon as he had obtained access to the instrument, he contrived to damage the stop imitating the human voice, which he had invented, and which had been its great merit, and thus punished the pride and cruelty of the municipality. In the remarkable Gothic Church of St. Peter is a chapel, built in commemoration of the infant St. Simeon, or Simonin, whose alleged martyrdom at the hands of the Jews, in 1472, I have already had reason to mention. Many relics of him are shown in the chapel, where a festa is still kept in his honour on March 24. The cutting of his name in the stone is still quite legible.

[1] A variant of this tradition takes the more usual form of applying it to the architect of the edifice, as with the Kremlin. As Stöber gives it from Strasburg, it was there the maker of the great clock.

My limits forbid my speaking in much detail of the secular buildings and institutions which are, however, not unworthy of attention. There are clubs and reading-rooms — in some of which aspirations after union with Italy are steadily propagated. The spirit of loyalty to Austria, though still strong in many breasts, has nothing like the same influence as in 1848–9, or in 1866, when the attacks and blandishments of the revolutionists of Italy were alike powerless to shake the allegiance of the Trentiners. No one will overlook the vast Schloss buon Consiglio in the Piazza d'Armi, said to be an Etruscan foundation. The public museum is a very creditable institution, enriched in 1846 by the legacy of Count Giovanelli's collection, chiefly of coins and medals; and paintings, not to be despised, are to be seen in the collections of the best families of the place—Palazzi Wolkenstein and Sizzo, Case Salvetti and Gaudenti. Two great ornaments of the city are the Palazzi Tabarelli, and Zambelli or *Teufelspalast*; and with the legend of the latter I must wind up my notice of Trent.

Georg Fugger, a scion of the wealthy Anthony Fugger, of Augsburg, the entertainer of Charles Quint, was deeply enamoured of the spirited Claudia Porticelli, the acknowleged beauty of Trent. Claudia did not appear at all averse from the match, but she was too proud to yield herself all too readily; and besides, was genuinely possessed with the spirit of patriotism, to which mountain folk are never wanting. Accordingly, when the reply long pressed for from her lips came at last, it informed him that never would Claudia Porticelli of Tirolean Trent give her hand to one whose dwelling was afar from her native city; she wondered, indeed, that one who did not own so much as a little house to call a home in Trent, should imagine he possessed her sympathies. To another this answer would have amounted to a refusal, for it only wanted a day of the time already fixed, of long date beforehand, for the announcement of her final choice. But Georg Fugger, whose vast riches had long nursed him in the belief that '*money* maketh man,' and that nothing was denied to him, would not yield up a hope so dearly cherished as that of making Claudia Porticelli his wife. To his determined mind there was a way of doing everything a man was resolved to do. To build a house, however, in one night, and that a house worthy of being the home of his Claudia, when men should call her Claudia Fugger, was a serious matter indeed. No human hands could do the work, that was clear; he must have recourse to help from which a good Christian should shrink; but the case was desperate; he had no choice. Nevertheless, Georg Fugger had no mind to endanger his soul either. The game he had to play was to get the Evil One to build the house, but also to guard from letting him gain any spiritual advantage against him; and his indomitable energy devised the means of securing the one and preventing the other. Without loss of time the devil was summoned, and the task of building the desired palace propounded. The tempter willingly accepted the undertaking, on his usual condition of the surrender of the soul of him in whose favour it was performed. Georg Fugger cheerfully signed the bond with his blood, only stipulating first for the insertion of one slight condition on his side—namely, that the devil should do one little other thing for him before he claimed his terrible guerdon. 'Whatever you like! it won't be too hard for me!' boasted the Evil One; and they separated, each well satisfied with the compact.

'The Devil's Palace has a splendid design, worthy the genius of Palladio,' writes a modern traveller, who has only seen it in its decadence. On the night in which it was built, it was resplendent with marbles and gilding and tasteful decoration; furnished it was too, to satisfy the most fastidious taste, and the requirements of the most luxurious. With pride the devil called Georg Fugger to come and survey the lordly edifice, and name his 'final condition.' Georg Fugger was prepared for him; he had taken a bushel of corn, and strewn it over all the floors of the vast building. 'Look here, Meister,' he said. 'If you can gather this corn up grain by grain, and deliver me back the whole number correctly, then indeed my soul will be yours; but if otherwise, my soul remains my own and the palace too. That is my final condition.'

The devil accepted he task readily, and with no misgiving of his success. True, it took all the time that remained before sunrise to collect all the scattered grain; still he had performed harder feats before that day. But the hours ran by, and still there were five grains wanting to complete the count; where could those five grains be! With a flaring torch, lighted at his fiercest fire, he searched every corner through and through, but the five grains were nowhere to be seen, and daylight began to appear! 'Ah! the measure is well-heaped up, the Fugger won't discover they are missing,' so the fiend flattered himself. But Georg Fugger was keener than he seemed. Before his eyes he counted out the corn, and asked for the five missing grains. 'Stuff! the measure is piled up full enough, I can't be so particular as all that. The number must be there.' 'But it is not!' urged Fugger. 'Oh, you've miscounted,' rejoined the Evil One; I'm not going to be put off in that way. I've built your house, and I've collected your measure of corn, and your soul is mine; you can't prove that there were five more grains.' 'Yes, I can,' replied Fugger; 'reach out me your paw;' and the Devil, not guessing how he could convict him by that means, held out his great paw, with insolent confidence of manner. 'There!' cried Fugger, pointing to it as he spoke; 'there, under your own claws, lie the five grains! That corn had been offered before the Holy Rood, and by the power of the five Sacred Wounds it was kept from fulfilling your fell purpose. You had not collected the full number of grains into the measure by the morning light, so our bargain is at an end. Begone!' The Devil, self-convicted, had no refuge but to strive to alarm his victor by a show of fury, and with burning claw he began tearing down the wall so lately raised. But Fugger remained imperturbable, for he had fairly won the palace, and the Devil himself had no more power over it. He could only succeed in making a hole big enough for himself to escape by, which hole was for many and many years pointed out.

But Fugger had also hereby established his claim to Claudia's hand, who rejoiced at the gentle violence thus done her; and many happy days they spent together in the *Teufelspalast*. In later years it passed from their family into the hands of Field-Marshal 'Gallas, who lived here in peaceful retirement after his renowned exploits in the Thirty Years' War, whence it was long called Palazzo Gallas or Golassi; but it has lately again changed hands, and thus acquired the name of Palazzo Zambelli.

The suburbs of Trent, among other excursions, offer the pleasing pilgrimage of the *Madonna alle Laste*,[1] which is reached through the Porta dell' Aquila, on the east side of the city, by half an hour's climbing up a mountain path off the road to Bassano. On a spur of this declivity had stood from time immemorial a marble *Maria-Bild*, honoured by the veneration of the people. Somewhere about the year 1630 a Jew wantonly disfigured and damaged the sacred token, to the indignation of the whole neighbourhood. Christopher Detscher, a

---

[1] Laste is dialectic for a smooth, steep, almost inaccessible chalk cliff.

German artist, devoted himself to restoring it; but it was impossible altogether to obliterate the traces of the injury. By some means or other, however—the people said by miraculous intervention—it was altogether renewed in one night; and this prodigy so enhanced its fame, that there was no case so desperate but they believed it must obtain relief when pleaded for at such a shrine. A poor cowherd named Antonia, who had been deaf all her life, was said to have received the power of hearing after praying there; and a child, who had died before there was time to baptise it, a reprieve of existence long enough to receive that Sacrament. The grateful people now immediately set themselves to raise a stone chapel over it, and by their ready alms maintained a hermit on the spot to guard the sacred precincts. Twelve years later, by the bounty of Field-Marshal Gallas, a community of Carmelites was established on the spot, which continued to flourish down to the secularization of Joseph II. The convent buildings, however, yet serve the beneficent purpose of a Refuge for foundlings and orphans. The prospect from the precincts of the institution is very fine; between the distant ranges of mountains and the foreground slopes covered with peach trees, lies the grand old city of Trent, shaped, like the country of Tirol itself, in the form of a heart.[1] Very effective in accentuating the outline

---

[1] Hence Kaiser Max was wont to call Tirol 'the heart' and 'the shield' of his empire.

are the two old castles of the Buon' Consiglio and the Palazzo degli Alberi, both formerly fortress-residences of the Prince-Bishops of Trent, the former vieing with the castle of the Prince-Bishop of Salzburg in extent and grandeur. The curious isolated rock of Dos Trento is another centre of a splendid view. The Romans called it Verruca, a wart. It was strongly fortified by Augustus, and remains of inscriptions and bas-reliefs are built into the wall of the ancient church of St. Apollinaria, occupying the site of a temple of Saturn. The vantage ground it afforded in repelling the entry of the French in 1703 obtained for it the name of the *Franzosenbühel*. It has lately been newly fortified. A charming but somewhat adventurous excursion may be made on foot, by a path starting from the fort of the Dos Trento rock, to the cascade of Sardagna. Somewhere about this path, in the neighbourhood of Cadine, it is said, St. Ingenuin,[1] one of the early evangelizers of the country, planted a beautiful garden, which was a living model of the Garden of Eden; but so divinely beautiful was it, that to no mortal was it given to find it. Only the holy Albuin obtained by his prayers permission once to find entrance

---

[1] St. Ingenuin was Bishop of Säben or Seben, A.D. 585. The See, founded by St. Cassian, had been long vacant, and great errors and abuses had taken root among the people, who in some places had relapsed towards heathen customs. His success in reforming the manners of his flock was most extraordinary. He built a cathedral at Seben, where he is honoured on Feburary 5, the anniversary of his death. St. Albuin, one of his successors, was a scion of one of the noblest families of Tirol; he removed the See to Brixen, A.D. 1004.

to 'St. Ingenuin's Garden.' Entranced with the delights of the place, he determined at least to bring back some sample of its produce. So he gathered some of its golden fruits, to show the children of earth. To this day a choice yellow apple, something like our golden pippin, grown in the neighbourhood, goes by the name of St. Albuin's apple.

The only remaining towns of any note in the line of the Wälschtirolische Etschthal, are Lavis and S. Michel. Lavis is a pretty little well-built town (situated at the point where the torrents of the Cembra, Fleims, and Fassa valleys, under the name of the Avisio, are poured into the Etsch), remarkable for a red stone viaduct, nearly 3,000 feet long, near the railway station, over the Avisio. Lavis fell into possession of the French in 1796, when the church was burnt and the houses plundered. In 1841—forty-five years after—a French soldier sent a sum of one hundred gulden to the church, in reparation for having carried off a silver sanct-lamp for his share of the booty.

Lavis has on many another occasion stood the early brunt of the attacks of Tirol's foes, and its people have testified their full share of loyalty. There is a tradition that the French, having on one occasion gained possession of it with a band two hundred strong, the people posted themselves on the neighbouring heights and harassed them in flank; but a cobbler of Lavis, indignant at the havoc the French were making, left this vantage ground, and running down into the town, shouting 'Follow me, boys!' dispersed the French troops before one of his fellows had time to come up![1]

San Michel, or Wälsch Michel, is the boundary town against the circle of South Tirol, once the last town on Venetian territory. There are imposing remains here of a fine Augustinian priory, which originated in a castle given up to this object by Ulrich Count of Eppan in 1143; the building has of late years been sadly neglected; it is now a school of agriculture. A little way before Wälsch Michel, the railway crosses, for the first time since leaving Verona, to the left bank of the Adige, by a handsome bridge called by the people 'the *sechsmillionen Brücke*.' Here we leave the Etschthal for a time, but we shall renew acquaintance with it in its northern stretch when we come to visit South Tirol.

The two northern tributary valleys of the Etschthal on the west are the Val di Non [2] and Val di Sole; among the Germans, they go by the names of Nonsberg and Sulzberg, as if they considered the hills in their case more striking than the valleys. The Val di Non is entered at Wälschmetz or Mezzo Lombardo by the strangely wild and gloomy Rochettapass. Wälschmetz is a flourishing Italian-looking town, whence a stellwagen meets every train stopping at

[1] This is a local application of the widsepread myth of the tailor. who kills 'seven at one blow,' identified by Vonbun (p. 71-2) with the *Sage* of Siegfried. Prof. Zarncke has also written a great deal to show Tirol's place in the Nibelungenlied.

[2] Anciently Anaunium, and still by local scholars called Annaunia, a possession of the Nonia family, not unknown to Roman history.

San Michel. Conveyances for exploring the valleys can be hired either at the 'Corona' or the 'Rosa.' The Rochetta is guarded by a ruined fort fantastically perched on an isolated spur of rock called Visiaun or Il Visione, said to have formed part of a system of telegraphic communication established by the Romans.

In the church of Spaur Maggiore, or Spor, so called because the principal place in the neighbourhood, which at one time all belonged to the Counts of Spaur, is a *Wunderbild* of the Blessed Virgin, which has for centuries attracted pilgrims from the whole country round. The church of the next place of any importance, Denno, is remarkably rich in marbles, and handsome for its situation; a new altar-piece of some pretension, and a new presbitery, were completed here in August 1869. Flavone or Pflaun, the next village, is particularly proud of a rich silver-gilt cross, twenty-five pounds in weight, and set with pearls, a gift of a bishop of Trient. At the time of the French invasion it was taken to Vicenza, but as soon as peace and security were re-established the people would not rest till it was restored to them. The hamlet is adorned with a rather handsome municipal palazzo, built in the sixteenth century, when the ancient Schloss, which overhangs the Trisenega torrent, was pronounced unsafe after several earth-slips. This valley is, if possible, richer in such remains than any other: every mountain spur bristles with them. One of the most important and picturesque is the Schloss Belasis, near Denno, claiming to be the cradle of the family of that name, which has established itself with honour in several countries of Europe, including our own. Behind Pflaun are large forests, which constitute the riches of the higher, as the *Seidenbaum* [1] is of the lower, level of the valley. In its midst lies the *Wildsee* of Tobel, which, frozen in winter, serves for the transport of the timber growing on the further side. The safety of its condition for the purpose is ascertained by observing the time when the trace of the sagacious fox shows that he has trusted himself across.

Cles, situated nearly at the northernmost reach of the valley, is a centre of the silk trade, and the factory-girls are remarkable for their tastefully adorned hair, though they all go barefooted. The site of a temple of Saturn, of considerable dimensions, has been found, coinciding with traditions of his worship having been popular here; and remains of an ancient civilization are continually dug up. There is a wild-looking plain outside the town, still called the *Schwarzen Felder*, or black fields, because tradition declares it to be the place where the Roman inhabitants burnt their dead. Here SS. Sisinius, Martyrius, and Alexander, are believed to have suffered death by fire on May 29, 397, because these zealous supporters and missionaries of St. Vigilius refused to take part in a heathen festival. St. Vigilius

---

[1] The white mulberry, whose leaves feed the silkworm, rearing which forms one great industry of Wälsch-Tirol, is called the *Seidenbaum*, the silk tree.

no sooner heard of their steadfast witnessing to the truth, than he repaired to the spot, and after zealously collecting and venerating their remains, preached so powerfully on their holy example, that great numbers were converted by his word. A church was shortly after built here, and being the first in the neighbourhood, was called *Ecclesia*, whence the name of Cles. The devout spirit of these saintly guides does not seem wanting to the present inhabitants; when the jubilee was held on occasion of the Vatican Council, more than two thousand persons went to Communion. At the not far distant village of Livo, on the same occasion, it was found necessary to erect a temporary building to supplement the large parish church, for the numbers who flocked in from the outlying parishes. The same thing occurred when the faithful were invited to join in prayers for the Pope after the Piedmontese invasion of Rome, September 20, 1870.

On these 'Campi neri' was found, in the spring of 1869, a tablet since known as the 'Tavola Clesiana.' It is a thickish bronze tablet, about 18 in. by 13 in., with holes showing where it was attached to a wall by the corners. It bears an inscription in Roman character, the graving of which is quite distinct and unworn, as if newly executed. It is as follows, and has given rise to a great deal of controversy among archæologists, and between Professors Vallaury and Mommsen, concerning its bearing on the early history of Annauria:—

Miunio . sIlano . q . sulpicio . camerino . CoS
idibus . martIs.   baIs . in . praetorio . edictum .
ti . claudi . caesaris . augusti . germanici . propositum .
                        fuit . id .
quod . infra . scriptum . est .
ti . claudius . caesar . augustus . germanicus . pont .
                        maxim .
trib . potest . VI . imp . XI . P . P . cos . designatus .
                        IIII . dicit .
cum . ex . veteribus . controversIs . petentibus . aliquamdiu .
                        etiam .
temporibus . ti . caesaris . patrui . meI . ad . quas .
                        ordinandas . pinarium .
apollinarem . miserat . quae . tantum . modo . inter .
                        comenses . essent .
quantum . memoria . refero . et . bergaleos . is que .
                        primum . apsentia .
pertinaci . patrui . meI .
deinde . etiam . gaI . principatu . quod . ab . eo . non .
                        exigebatur .
referre . non . stulte . quidem . neglexerit . et . posteac .
                        detulerit . camurius .
statutus . ad . me . agros . plerosque . et . saltus . meI .
                        iuris . esse . in . rem .
praesentem . mIsi . plantam . iulium . amicum . et .
                        comitem . meum . qui .
cum . adhibitis . procuratoribus . meis . quisque . in .
                        alia . regione .
quique . in . vicinia . erant . summa . cura . inquisierit .
                        et . cognoverit .
cetera . quidem . ut . michi . demonstrata . commentario .
                        facto . ab . ipso . sunt .
statuat . pronuntietque . ipsi . permitto .
Quod . ad . condicionem . anaunorum . et . tulliassium .
                        et . sindunorum .
pertinet . quorum . partem . delator . adtributam
                        tridentinis .
partem . neadtributam . quidem . arguisse . dicitur . tam .
                        et . si .
animaduerto . nonnimium . firmam . id . genus . hominum .
                        habere . civitatis .

romanae . originem . tamen . cum . longa . usurpatione .
in . possessionem .
eius . fuisse . dicatur . et . ita . permixtum . cum .
tridentinis . ut . diduci .
ab . Is . sine . gravi . splendi . municipI . iniuria . non .
possit . patior . eos .
in . eo . iure in . quo . esse . existimaverunt . permanere .
beneficio . meo .
eo . quidem . libentius . quod . plerisque . ex . eo
genere . hominum . etiam .
militare . in . praetorio . meo . dicuntur . quidam . vero .
ordines . quoque .
duxisse . nonnulli . collecti . in . decurias . romae . res .
iudicare .
Quod . beneficium . Is . ita . tribuo . ut . quaecumque
tanquam . cives .
romani . gesserunt . egeruntque . aut . inter . se . aut .
cum . tridentinis .
alIsve . ratam . esse . iubeat . nominaque . ea . que .
habuerunt . antea .
tanquam . cives . romani . ita . habere . Is . permittam .

A fragment of an altar was found at the same time, with the following words on it :—

SATURNO SAC[R]
L. PAPIRIUS L
OPUS

Livo is the first village of the Val di Sole, which runs in a south-westerly direction, forming nearly a right-angle with the Val di Non, than which it is wilder, and colder, and less inhabited. At Magras the Val di Rabbi strikes off to the north. Its baths are much frequented, and S. Bernardo is hence provided with four or five capacious hotels. A new church has just been built there, circular in form, with three altars, one of which is dedicated in honour of St. Charles

Borromeo, who visited the place in 1583, and preached with so much fervour as effectually to arrest the Zuinglian teaching, which had lately been imported.

Male is the chief place of Val di Sole, and contains about 1,500 inhabitants. At a retreat held here last Christmas by the Dean of Cles, so many of them as well as of the circumjacent hamlets were attracted, that not less than 3,000 went to communion. Further along the valley is Mezzana, the birthplace of Antonio Maturi, who, after serving in the campaigns of Prince Eugene, entered a Franciscan convent at Trent, whence he was sent as a missionary to Constantinople, and was made Bishop of Syra, and afterwards was employed as nuncio by Benedict XIV. It was almost entirely destroyed by fire a few years ago, but is being rapidly rebuilt. After this place the country becomes more smiling, and cheerful cottages are seen by the wayside, with an occasional edifice, whose solid stone-built walls suggest that it is the residence of some substantial proprietor. The valley widens out to a plain at Pellizano, round which lofty mountains rise on every side. The church here has a most singular fresco on the exterior wall, which is intended to record the circumstance that Charles Quint passed through in 1515. Some restoration or addition was made to the church at his expense, and a quaint inscription hints that he did it somewhat grudgingly.

A few miles further the valley divides into two branches, the Val di Pejo and the Val di Vermiglio. At Cogolo, the chief place of Val di Pejo, had long been stored a magnificent monstrance, offered to the church by Count Migaczy, who, though resident in Hungary, owned it for his *Stammort*.[1] It had long been the admiration of the neighbourhood, and the envy of visitors; but it was stolen by sacrilegious hands in the troubles consequent on the invasion of the Trentino by 'Italianissimi,' in 1849. Count Guglielmo Megaczy sent the village a new one of considerable value and handsome design, whose reception was celebrated amid lights and flowers, ringing of bells and firing of *mortaletti*, July 18, 1869. This branch of the valley is closed in by the *Drei Herren Spitz*, or *Corno de' tre Signori*, the boundary-mark between the Valtellina, Bormio, and Tirol, and so called when they belonged to three different governments. The Val di Vermiglio is closed by Monte Tonale, the depression in whose slope forms the Tonal Pass into Val Camonica and the Bergamese territory. Monte Tonale was notorious in the sixteenth to early in the eighteenth century for its traditions of the Witches' Sabbath, and the trials for sorcery connected with them.[1] Freyenthurn, a ruin-crowned peak at no great distance, bears in its name a tradition of the worship of Freya.

On the vine-clad height of Ozolo, above Revo,

---

[1] *Stammort*, Cradle of his race.

[2] See *Un processo di Stregheria in Val Camonica*, by Gabriele Rosa, pp. 85, 92; and *Il vero nelle scienze occulte*, by the same author, p. 43; and Tartarotti *Congresso delle Lammie*, lib. ii. § iv. It is one of the only four such spots anywhere existing where Italian is spoken.

a few miles north of Cles, is a little village named Tregiovo, most commandingly situated; hence, on a fine day, may be obtained one of the most enchanting and remarkable views, sweeping right over the two valleys. Hence a path runs up the heights, and along due north past Cloz and Arz to Castelfondo, with its two castles overhanging the roaring cascade of the Noce. Along this path, where it follows the Novella torrent, numbers of pilgrims pass every year to one of the most famed sanctuaries of Tirol—*Unsere liebe Frau im Walde,* or *auf dem Gampen,* as the mountain on which it is perched is called by the Germans; and this reach of the Nonstal is almost entirely inhabited by Germans. The Italians call it *le Pallade,* and more commonly *Senale.* The chapel is on the site of an ancient hospice for travellers, which became disused, however, as early as the fourteenth century. A highly-prized *Madonnabild,* of great sweetness of expression, found in a swamp near the place, stands over the high-altar. A celebration of the seventh centenary of its being found was kept by a festival of three days from August 14, 1869, when crowds of pilgrimages, comprising whole populations of circumjacent villages, both German and Italian, might have been seen gathering round the shrine. Fondo, though but a few miles distant, is a thoroughly Italian town; and so great is the barrier this difference of tongue sets up, that great part of the population of the one never visits the other. It was nearly burnt down in 1865, and has hardly yet recovered from the catastrophe; the church, which occupies a very commanding situation, was saved, and its fine peal of six bells. Near it is St. Biagio, where was once the only convent the Nonsthal ever possessed. Near this again is Sanzeno, which, by a tradition a little different from that given at Deuno, is made out to be the place of martyrdom of SS. Sisinius (supposed to be another form of the name of St. Zeno), Martyrius, and Alexander. Their relics, at all events, are venerated here in a marble urn behind the high-altar of the church, which bears the title of the Cathedral of the Val de Non; and the Roman remains, which are continually being discovered,[1] show that there were Romans here to have done the martyrdom. The legend is, that these saints were three brothers of noble family, of Cappadocia, who put themselves under the bidding of S. Vigilius, Bishop of Trent (who was already engaged in the conversion of the valley), A.D. 390. Their conversions were numerous during a series of years; but on May 23, 397, the inhabitants of the valley, who adhered to the old teaching, desirous to make their usual sacrifice to obtain a blessing on their crops, called upon the Christian converts to contribute a sheep for the purpose. On the Christians refusing a strife ensued, of which two of the three missionaries were the immediate victims; but the next day, the

---

[1] A mithraic sacrifice with several figures, sculptured in bas-relief. in white Carrara marble, in very perfect preservation, bearing the inscription:

ILDA MARIVS
L. P.

has just been found at this very spot.

third, Alexander was also arrested; he was burnt alive, along with the corpses of his companions. A church was subsequently built on the spot where they were said to have suffered; their acts may be seen in a bas-relief of the seventeenth century. San Zeno is also famous for being the birth-place of Christopher Busetti, whose verses, no less than the details of his life, earned for him the title of the Tirolean Petrarch. A little east of San Zeno is the narrow inlet into the Romediusthal, so called from S. Romedius, whom we heard of at Taur,[1] having chosen it for a hermitage whence to evangelize the Nonsthal, and in which to end his days. A more secluded spot could not be found on the whole earth. Perpendicular rocks narrow it in, leaving scarcely a glimpse of the sky above; the torrent which files its way through it, called San Romedius-Bach, continually works a deeper and deeper bed. Two other torrents strive for possession of the gorge (Romediusschlucht), the Rufreddo and the Verdes, between them; near their confluence rises a stark isolated crag, from whose highest point, almost like a fortress, rises the farfamed hermitage, accessible only from one side. The legend has it that S. Vigilius, knowing his exalted piety, conceived the idea of consecrating the cell whence his holy prayers had been poured out, for a chapel, but was warned in a vision that angels had already fulfilled the sacred task. When this was known, it may be imagined that the veneration of the people for it knew no bounds, and the angelic consecration is still remem-

---

[1] See pp. 164-6.

bered by diligent pilgrimages every first Sunday in June; the Saint's feast is on January 15. The shrine is overladen with thank-offerings, which might attract the robber in so lonely a situation. Due precautions are taken for the preservation of the treasury; the chapel is surrounded by strong walls, and ingress is not permitted to strangers after nightfall. There is no record of any attempt having been made on it but once, some thirty years ago. On this occasion three men presented themselves at the gate, and urgently begged to be admitted to confession; their devotion was so well assumed, and their show of penitence so hearty, that the good priest could not refrain from letting them in. He had scarcely taken his seat in the confessional, however, than the three surrounded him, each presenting a pistol at his breast; all three missed fire, and the would-be robbers, convicted by the portent, knelt and made a real confession of their misdeeds, and left as really penitent as they had feigned to to be on arriving.

The spot has never ceased to be honoured since the death of the saint, somewhere about 398. It is strange to stand between the walls of the living mountain and realize the fact. There are few shrines in all Europe which can boast of such antiquity, such unbroken tradition, and such exemption from desecration. The building is as singular and characteristic as the locality. The chapel, where the saint's remains rest, and where he himself raised the first sanctuary of the Nonsthal, is reached by one hundred and twenty-two steps, necessarily very steep; and on attaining the last, it must be a very steady head that can turn to survey the rise without giddiness. The interior is quite in keeping with the surroundings. Its light is dim and subdued, sufficient only to reveal the countless trophies of answered prayer which cover the dark red marble columns and enrichments. There are two other chapels at lower levels, one of the Blessed Sacrament, called *del Santissimo*, and one over the hermitage in the rock. Flanking this curious pile of chapels on chapels are, on one side, the priory or residence of the chaplain of the place, and on the other the Hospice for pilgrims and visitors, the whole forming a considerable *corps de bâtiment*, and enclosed by a wall which seems to have grown out of the rock. Another little crag, jutting up as if in emulation of that so gloriously crowned, was made into a *Gottesacker*, by a late prior, and its churchyard cross affords it a striking termination too; though not many monuments of the dead bristle from its sides as yet. This singularly interesting excursion may be made direct from S. Michel by those who have not time for visiting the whole valley. They will pass several striking old castles, particularly that of Thun, nearly opposite Castle Bellasi, the *Stammschloss* of one of the oldest and noblest German families, founded by one of the dearest companions and patrons of St. Vigilius. No other has given so many distinguished scions to the service of the Church; Sigmund von Thun was the representative of the Emperor at the Council of Trent. There is a strong attachment between it and the people of the valley, who delight in celebrating every domestic event by what they call a Nonesade, or poem in the dialect of the Val di Non. The castle is well kept up; the interior is characteristically decorated and arranged, and many curiosities are preserved in the library; its grounds also are charmingly laid out. It is supplied with water by a noble aqueduct, raised in 1548, right across the valley from Berg St. Peter; crowned also by an ancient castle, but in ruins. Few will have a prettier page in their sketch-book than they can supply it with here.

Half way between Sanzeno and Fondo, by a path which forms a loop with that already mentioned, by Cloz and Arz, and just where the opening into the Romediusthal strikes off, is a village named Dambel or Dambl, where a very curious relic of antiquity, and an important one for throwing light on the history of the earlier inhabitants of the valley, was unearthed a couple of years ago. It is a stout, handsome bronze key, $14\frac{1}{2}$ in. long, the bow ornamented with scroll-work, which at first sight suggested the idea that it had formed part of a comparatively modern casting of the Pontifical arms. Closer inspection showed that on an octagonal ornament of the upper part of the stem was an inscription, not merely engraved, but deeply cut (it is thought with a chisel), and in perfect preservation, in characters described by a local antiquary as ' parte

Runiche, parte Gotiche, del Greco e Latino del 388 dell' era volgare, descritte da Ufila; ma molte somigliano a quelle del Latino dell' Ionio 741 B. C."

The owner of the ground, Bartolo Pittschneider, the jeweller of the village, seems to have been digging the foundation for a rustic house, intending to make use of a remnant of a very ancient wall long thought to have formed part of a temple of Saturn. At a depth of about 18 or 20 in. he came to a sort of pavement, or tomb or cellar covering, of roughly-shaped stones resting against and sloping away from the base of the ancient wall, so as to form a little enclosure. Along with the key lay some other small objects, which unfortunately have been dispersed,[1] but among them were two bronze coins of Maximinian and Constantine the Great, thought to indicate the date of the burial of the key and not that of its manufacture.

This key was subsequently sent to Padre Tarquini,[2] and a copy has been given me of his report upon it. He pronounced the inscription to be undoubtedly Etruscan, but at the same time he did not think

[1] Too many such remnants, which the plough and the builder's pick are continually unearthing, have been thus dispersed. It has been the favourite work of Monsignor Zanelli, of Trent, to stir up the local authorities to take account of such things, and so form a museum with them in Trent.

[2] Padre Tarquini—one of the rare instances of a Jesuit being made a Cardinal—died, it may be remembered, in February last, only about two months after his elevation. He had devoted much time to the study of Etruscan antiquities; he published *The Mysteries of the Etruscan Language Unveiled* in 1857, and later a Grammar of the language of the Etruscans.

the work of the key to be of older date than the fourth century of our era; inasmuch as there are other examples of Etruscan writing surviving to as late a date in remote districts; that its size and material (a mixture of silver and copper) denoted it to belong to some important edifice, and most probably to the very temple of Saturn amid whose ruins it was found buried. He found in it two new forms of letters not found in other Etruscan inscriptions, but says that similar aberrations are too common to excite surprise. He translated it in the following form:—' Ad introducendum virum (1) addictum igni in Vulcani (2) Vivus aduratur ob perversitatem—incidendo incide (3)—Sceleratus est; sectam facit; blasphemavit—In aspectu ejus ascendentes limen paveant, videntes hominem oblitum Ejus (4) præstare jubilationem retinenti ad cruciatum, tamquam hostem suum.'

It would be curious to know how Mr. Isaac Taylor would read the inscription by his different method, for Padre Tarquini found a curious coincidence of circumstances to afford an interpretation to his translation. It would seem that it was only after translating it as above that his attention was called to the Christian

'(1.) Or it might be 'ad introductionem viri.' (2.) 'Vulcano' here (precisely as in another Etruscan inscription found a few years before at Cembra, and translated by Professor Giovanelli) for 'ignis.' (3.) An allusion to the custom of first piercing (sforacchiare) the bodies of persons to be burnt in sacrifice, which appears from the inscription found at S. Manno, near Perugia, and again from the appearance of the figures of human victims represented in the Tomba Vulcente. (4.) The deity of the place to which the key belonged, probably, therefore, Saturn.'

local tradition, and then he was struck with several points of contact between it and them. 1. The date which he had already assigned to the key is that given by the Bollandists to the martyrdom of St. Alexander and his two brothers. 2. It was found within the very precincts where he was said to have been burnt, and (his translation of) the inscription commemorates a human burnt sacrifice (*il vivicomburio*). 3. The inscription (by his translation) seems to allude to Christians, to their suffering expressly for propagating their religion. 4. The inscription points to the sacrifice having taken place in an elevated situation, as it uses the verb 'to ascend,' and the contemporary narrative of St. Vigilius to St. Chrysostom of the event, as it had happened before his eyes, says 'Itum est post hæc in religiosa fastigia, hoc est altum Dei templum . . . in conspectu Saturni.' He further goes on to approve a conjecture of the local antiquary that the key was a votive offering made on occasion of the martyrdom of St. Alexander with SS. Zeno and Martyrius, in thanksgiving for the triumph over their teaching, and inscribed with the above lines as a perpetual warning to their followers.

The Avisiothal—the northernmost eastern tributary of the Etschthal—consists of three valleys running into each other; the Val di Cembra, or Zimmerthal; the Val Fieme, or Fleimserthal; and the Val di Fassa, or Evasthal. The Val di Cembra is throughout impracticable for all wheeled traffic. Nature has made various rents and ledges in its porphyry sides, of which hardy settlers have taken advantage for planting their villages, and for climbing from one to another; but even their laborious energy has not sufficed to make roads over such a surface. This difficulty of access has not been without its effect in tending to keep up the honesty, hospitality, and piety of the people; but as few will be able to penetrate their recesses, their characteristics will be better sacrificed to the exigencies of space than those of others. I will only mention, therefore, the Church of Cembra, the *Hauptort* (about four hours' rugged walk from Lavis), which is an ancient Gothic structure well kept up, and adorned with paintings; and a peculiar festival which was celebrated on the Assumption-day, 1870, at Altrei, namely, the presentation of new colours to the Schiess-stand, by Karl von Hofer, on behalf of the Empress of Austria. One bears a Madonna, designed by Jele of Innsbruck, on a banner of green and white (the national colours); the other the names of the Empress ('Karolina Augusta') and the word 'All-treu,' the original name of the village, conferred on it by Henry Duke of Bohemia, when he permitted ten faithful soldiers to make a settlement here free of all taxes and customs. And yet the Italians, regardless of derivations, have made of it Anterivo.

Cavalese (which can be reached in five hours by *stellwagen* running twice a day from the railway station at Neumarkt) stands near the point where the Val di

Cembra (which runs nearly parallel to the railway between Lavis and Neumarkt) passes into the Fleimserthal. It is a charmingly picturesque, thriving little town, and should not be overlooked, for the church is a very museum of Tirolese art: painting, sculpture, and architecture, all being due to native artists, and highly creditable to national taste, culture, and devotion. Among these artists were Franz Unterberger, who was chosen by the Empress Catherine to execute copies from Raffael's *Loggie,* Alberti, Riccaboni, and others, whose fame has resounded beyond the echoes of their native mountains. Many private houses also contain works of Tirolese art. Cavalese stands on a plateau, overlooking a magnificent panorama, and shaded by a grove of leafy limes. Under these is a stone table, with stone seats arranged round it, where a sort of local parliament was formerly held. Respecting the appropriation of this plateau for the site of the church, tradition says that in early times, when the church was about to be built, the commune fixed upon this plateau, in the outskirts of the town, as the most beautiful, and therefore most appropriate, situation. But the old lady, part of whose holding it formed, could be induced on no consideration to give it up. Some little time after, however, she had a very serious illness; on her sick bed she vowed, that if restored to health she would devote as much of her fair meadow to the use of the church as a man could mow in one day.[1] She had no sooner registered her vow than health returned. The commune appointed a mower, and he mowed off the whole of the vast meadow in one day. The old lady always maintained that there was something uncanny about it, and anyone can see for themselves that no human mower could have done it. The Market-place is adorned with a very handsome tower. A new church is now building, after the design of Staidl, of Innsbruck, on the site of the little ruined church of St. Sebastian, which shows that the study of architecture is not neglected in Tirol. The space being very restricted, the novel expedient has been resorted to of placing the sacristy *under* the sanctuary, and with good effect to the external appearance. The former palace of the Bishops of Trent, now a prison, is not to be overlooked. Predazzo is the only other spot in this valley we will stop to look at. The extraordinary geological formation of the neighbourhood has attracted many men of science to the place, whose names may be seen in the strangers' book. The people are singularly thrifty and industrious. A high road connecting it with Primiero is just completed, which is to be continued to meet the railway projected between Belluno and Treviso. A new church is being raised there, of proportions and design quite remarkable for so remote a place. It was begun simultaneously with the

---

[1] A *Tag-mahd*, or 'day's mowing,' is a regular land measure in North Tirol.

troubles in Italy, in 1866, and a creditable amount has been since laid out upon it. The lofty vaulting of the nave is supported by ten monolithic columns of granite; the floor is paved with hard cement, arranged in patterns formed in colour; the smaller pillars, doors, steps, mouldings, are all of granite; much of the tracery is very artistic; the windows are of creditable painted glass, though not free from the German vice of over-shading. The architect is Michel Maier, of Trent; the elegant campanile by Geppert, of Innsbruck. It will be the largest church in the whole of Wälsch-Tirol, after the Cathedral of Trent. The interior arrangements and decoration bid fair to be worthy of the structure. There is some good polychrome in the presbytery, by Ciochetti, a young artist, native of the village of Moena, in Fassathal, who in the last five years has had eleven medals from the Academy of Fine Arts at Venice. It is the custom all through the valley that each village should have its own gay banner, which is carried before bridal processions to and from the church. But at Predazzo they have many other peculiarities; among these is the following :—The night before the wedding the bridegroom goes to the house of the bride, accompanied by a party of musicians, knocks at the door, and demands his bride. The eldest and least well-favoured member of the household is then brought to him, on which a humorous altercation takes place and a less ancient dame is brought, and so on, till all have been passed in review, and then the intended bride herself is brought at last, who admits the swain to the evening meal of the family. The friends and neighbours then come in, and bring their wedding gifts to the loving pair.

The Fassathal begins just after Moena. One of its wildest legends is that of the *feuriger Verräther*. It dates from the time of the Roman invasion. The mountain-dwellers appear to have been as zealous defenders of their native fastnesses then as in later times, and it is said the conquering legions were long wandering round the confines without finding any who would lead them into the interior of the country. It was at last an inhabitant of the Fassathal who betrayed the narrow pass which was the key to their defences, and which cost the liberty of the nation—all for the sake of the proffered blood-money. But he was never suffered to enjoy it; for a flash like lightning, though under a clear sky, struck him to the earth, and ever since, the traitor has been to be met by night wrapt in flames, and howling piteously.

Vigo is the principal town, and serves as the starting-point for the magnificent mountain excursions of the neighbourhood. The most difficult of these, and one only to be attempted by the well-seasoned Alpine climber,[1] is that of the massive snow-clad Marmolata, 10,400 feet high, surnamed the Queen of the Dolomites; but she is a severe and haughty queen, who knows

---

[1] There is no record of her summit ever having been attained before the successful ascent of Herr Grohmann, in 1864. Mr. Tucker, an Englishman, accomplished it the next year.

how to hold her own, and keep intruders at a distance; and many who have been enchanted with her stern beauty from afar have rued the attempt at intruding on the cold solitude of her eternal penance. For the legends tell that in her youth she was covered with verdant charms, which made her the delight of the people; but they were not content to use with pious moderation the precious gifts she had in store, and for some sin of theirs—some say for selfish disregard of the law of charity to the poor;[1] some say for disregard of the Church's law forbidding to work on the *hohe Unserfrauentag* (the Assumption),[2] some say for unjust

---

[1] I have given some of the most curious of these in a collection of *Household Stories from the Land of Hofer*.

[2] There is no tradition more universally spread over Tirol than that which tells of judgments falling on non-observers of days of rest. They are, however, by no means confined to Tirol. Ludovic Lalanne, *Curiosités des Traditions*, vol. iv. p. 136, says that the instances he had collected showed it was treated as a fault most grievous to heaven. 'Matthieu Paris, à l'année 1200, raconte qu'une pauvre blanchisseuse ayant osé travailler un jour de fête fut punie d'une étrange façon; un cochon de lait tout noir s'attacha à sa mamelle gauche.' He relates one or two other curious instances—one of a young girl who, having insisted on working on a holiday, somehow got the knot of her thread twisted into her tongue, and every attempt to remove it gave intolerable pain. Ultimately she was healed by praying at the Lady-altar at Noyon, and here the knot of thread was long shown in the sacristy.

I well remember the English counterpart in my own nursery. There were, indeed, two somewhat analogous stories; and I often wondered, without exactly daring to ask, why there was so much difference in the tone in which they were told, for the one seemed to me as good as the other. The first, which used to be treated as an utter imposture, was that a woman and her son surreptitiously obtained a consecrated wafer for purposes of incantation (we have had a Tirolean counterpart of this at Sistrans, *supra* pp. 221-2), and in pursuit of their weird operation had pierced it, when there flowed thereout such a prodigious stream of blood

striving for the possession of the soil—the vengeance of Heaven overtook them, and the once smiling meadows were converted into the hard and barren glacier. Near Vigo is a little way-side chapel, highly prized, because near it some French soldiers in the invasion of 1809 lost their way, and the town was thus saved from their depredations; and the legend arose that the Madonnabild had stricken them blind. Several of them died of falls and hunger, and tradition says, that on wild nights notes of distress from a dying bugler's horn may be heard resounding still.

The Avisio was once the boundary against Venetian territory; and St. Ulrich dying on its banks, on his return from Rome, exacted of his disciples a promise that they would carry his body across, so that he might find his final rest on German soil.

that the whole place was inundated, and all the people drowned. The second, which was told with something of seriousness in it, ('and they say, mind you, *that* actually happened,') was of a young lady who, having persisted in working on Sunday in spite of all her nurse's injunctions, pricked her finger. No one could stop the bleeding that ensued, and she bled to death for a judgment; and whether it was true or not, there was a monument to her in Westminster Abbey. Dean Stanley, who seems to have missed nothing that could possibly be said about the Abbey, finds place, I see, to notice even this tradition (pp. 219-20 and note), and identifies it with the monument of Elizabeth Russell (born 1575) in St. Edmund's Chapel. Madame Parkes-Belloc tells me she has often seen a wax figure of a lady (in the costume of two centuries later than Elizabeth Russell) under a glass case in Gosfield Hall, Essex (formerly a seat of the Buckingham family), of which a similar tradition is told.

CHAPTER XII.

WÄLSCH-TIROL.

VAL SUGANA.—GIUDICARIA.—FOLKLORE.

Legends are echoes of the great child-voices from the primitive world; so rich and sweet that their sound is gone out into all lands.

VAL SUGANA is watered by the Brenta through its whole course, running nearly direct east from Trent. It is reached by the Adler Thor, and over the handsome bridge of S. Ludovico, through luxuriant plantations of mulberries and vines, and with many a summer villa on either hand. The road leads (at a considerable and toilsome distance) to the low range of hills (in Tirol called a *Sonnenberg*) of Baselga, locally named Pinè, whose sides are studded with a number of villages and groups of houses. In one of these, Verda or Guarda by name, near the village of Montanaga, is the most celebrated pilgrimage of the Trentino—the Madonna di Pinè, also known as the Madonna di Caravaggio. It was the year 1729; a peasant girl, Domenika Targa, native of Verda, who was noted by all her neighbours for the angelic holiness of her life, had lost some of her herd upon the mountain one hot August day; in her distress, she knelt down to ask for help to bring back her charge faithfully. Suddenly the place was bathed in a light of glory, and before her stood a lady so benign and glorious, she could be none other than the *Himmelskönigin*. 'Go, my child, and tell them that you have seen me here, and that I have chosen this spot for my delight; and that their prayers will be heard which they offer before the picture of the Madonna di Caravaggio.' The light faded away, and Domenika turned to seek her flock. She found them all in order, waiting for her to drive them home. There was considerable discussion after this as to what 'Madonna di Caravaggio' might mean; and it was at last decided that it could mean nothing but the picture of the Madonna by Caldara, surnamed Caravaggio from his birthplace, venerated at Milan. Domenika could not leave her herds to go to Milan, and she was perplexed how to obey the vision. In her simple faith she addressed her prayer on high for further direction, and once more the heavenly sight was vouchsafed to her, and it was explained that the *Madonnabild* meant was not that of Milan, but the one in the little field-chapel of S. Anna, near Montanaga. Domenika did not fail to go there the next festival on which it was open, the Ascension Day, which was, that year, May 26. Above the faint light of the tapers tempered by the incense clouds, and amid the chanted litanies of the choir, the fair Queen once more appeared to her in garments of gold, and surrounded by a glittering train of attendants. Some months passed, and though the people had wondered at the marvel, nothing had been done to commemorate it; Domenika was kneeling, on September 8, the Nativity of the Blessed Virgin, in the Chapel of S. Anna. A sound of soft chanting broke on her ear, which she thought must be the procession of the parish coming up the hill to pray for rain. But as it grew nearer, the same heavenly radiance overspread the place, and once more she saw the Virgin Mother; but this time she looked stern, for the great favour of her visit had been overlooked, and she reasoned with Domenika on the ingratitude it betokened. Domenika honestly outspoke her inward cogitations on the subject—what could a poor cattle-herd do? It was given her to understand that much might be done even by such a poor peasant, if she exercised energy and devotion. With new strength and determination, she girt herself for the task of building a shrine over the spot so dear to her. At first she met with great ridicule and scorn, but she pursued her way so steadily and so humbly, that all were won to share her convictions. Offerings for the work began to flow in. Those who had no money gave their corn, or their grapes, their ornaments, and their very clothes. Year by year the new church rose, according as she could collect the means; and at last, on May 26, 1751, she had the consolation of seeing the complete edifice consecrated. It is a neat cruciform building, sixty-three feet long and fifty-three feet wide, with three marble altars, on one of which is a copy of the Madonna di Caravaggio of Miian painted by Jakob

Moser after he had made three pilgrimages to the original. I was not able to ascertain what was supposed to have been intended in the first instance by calling the old picture in S. Anna's field-chapel the Madonna di Caravaggio. Possibly the little Milanese town, which has given two painters to fame, had produced some 'mute inglorious' 'Caravaggio,' who painted the earlier picture. The commemoration of Domenika's vision is celebrated every year in Val Pinè by pilgrimages on May 26 when the most striking gatherings of Tirolese costume are to be observed there.

Pergine is the first large village on returning into the main valley, about six miles from Trent. It well deserves to be better known: the neighbourhood is of great beauty, and the form of the surrounding heights is well likened by the inhabitants to a theatre. The church, built in 1500–45, is spacious and handsome, adorned in the interior with red marble columns. In the churchyard are the remains of the older church, where every Lent German sermons are still preached for the benefit of the scattered German population, whose name for the place is Persen. The German and Italian elements within the village are blended with tolerable amity. From the fourteenth to the sixteenth centuries, silver, copper, lead, and iron, were got out in the neighbouring Fersinathal; and though the works are now nearly given up, the *Knappen* then formed an important portion of the community. They cast the bell as an offering to the church when building, and it is still called the *Knappinn*—by the Italians *canòppa*. The chief industry now is silk-spinning. The greatest ornament of the place is the *Schloss* of the Bishop of Trent, which is well kept up, and from the roof of which an incomparable view is obtained. Among the peculiar customs of the place those concerning marriages deserve to be recorded, as they tend to show the character of the people. Two young men of the bridegroom's friends are selected for the office of *Brumoli* so called; they have to carry, the one a barn-door fowl, the other a spinning-wheel, before the bride as she goes to and from church, to remind her of her household duties. After the wedding, as she returns with her husband to his house the door is suddenly closed as she approaches, and there is then carried on a dialogue, according to an established form, between her and her husband's mother—the latter requiring, and the former undertaking, that she will prove herself God-fearing and domesticated; that she will be faithful and devoted to her husband, and live in charity with all his family. The little ceremony complete, the mother-in-law throws wide the door, and receives her with open arms.

On the south side of the valley, opposite Pergine, is the clear lake of Caldonazzo, whose waters reflect the bright green chestnut woods around it; it is the source of the Brenta, and one of the largest lakes of Tirol; about three miles long, and half as broad. Count Welfersheim, an Austrian general, and his adjutant, were drowned in attempting to walk over the thin ice on it in March 1871. On a rugged promontory jutting into its midst stands the most ancient sanctuary of the neighbourhood, San Cristofero; once a temple to Saturn and Diana, but adopted for a Christian church by the earliest evangelizers of the valley, for which reason the produce of the soil and waters yet pays tithe to the presbytery of Pergine. Other villages add to the surrounding beauties of the lake, particularly Campolongo, with its church of St. Teresa high above the green waters, and the church and hermitage of San Valentin; the latter is now used for a *roccolo*, or *vogeltennen*, by which numbers of birds of passage are caught on their migrations. The land is very poor. To eke out their living, most of the male inhabitants of the villages around are wont to go out every winter as pedlars, with various small articles manufactured in the valley, and with which they are readily trusted by those who stay behind. On their return, which is always at Easter, they distribute honourably what they have earned for each, deducting a small commission. So straightforward and honourable are they, that though they have little idea of keeping accounts, and the sums are generally made out with a bit of chalk on the inn table, yet it is said that such a thing as a dispute over the amounts is utterly unknown. The church of St. Hermes, at Calzeranica, is reckoned the most ancient of the whole neighbourhood; remains of an ancient temple, thought to have been to

Diana of Antioch, have been found when repairing it. In the forest behind Bosentino, a neighbouring village, is a pilgrimage chapel called *Nossa Signora del feles; die h. Jungfrau vom Farrenkraut*—St. Mary of the Fern. Some two hundred years ago, Gianisello, a little dumb boy of Bosentino, who was minding his father's herd in the forest, was visited by a bright lady, who pointed to a tuft of fern growing under a chestnut tree, and bid him go and tell the village people she would have them built a chapel there. When the people heard the boy tell his story, who for all the twelve years of his life had never spoken a word before, they felt no doubt it was the Blessed Virgin he had seen. The chapel was soon built, and furnished with a painting embodying the little boy's story. In time of dearth, drought, epidemic, or other local calamity, many are the processions which may yet be seen wending their prayerful way to the chapel of St. Mary of the Fern.

Among the wild and beautiful legends of this part of the valley is a variant of one familiar in every land. A young swain, the maiden of whose choice was called to an early grave, went wandering through the chestnut groves calling for his beloved, till he grew weary with crying, and laid him down in a cave to rest. A sweet sleep visited him, and he found himself in it at home as of old in the Valle del Orco,[1] with his Filomena on his arm; he led her to the village church, and the

[1] It is significant of a symbolical intention that the story should

silver-haired pastor gave the marriage blessing, while all the village prayed around. He brought Filomena home to his old house, *alle Settepergole*,[1] his dear old father and mother welcomed her, and she brought sunshine into the cottage; and when they were called away the old walls were yet not without life and joy, for it resounded to the voice of the prattling little ones. The little ones grew up into stalwart lads and lasses, who earned homesteads of their own, and erewhile brought another tribe of prattling little ones to his knee; while Filomena smiled a bright sunshine over all, and they were so happy they prayed it might never end: but one day it seemed that the sunshine of Filomena's smile was not felt, for she was no longer there; then all grew pale and cold, and with a sudden chill he woke. It was grey morning as he rose from the cave; the cattle were lowing as they were led out to pasture; he looked out towards the chestnut groves, and watched in their waving foliage the strange effect which had been the charm of his childhood, looking like rippled ocean pouring abroad its flood.[2] But when he reached the village the sights and sounds were no more so familiar: the old church tower was capped with a

thus allude to the Valle del Orco; the more so as I cannot hear of any such actual locality in Val Sugana, though 'Orco' has lent his name to more than one spot, as we shall see later. There is, however, a Val d'Inferno between this valley and Predazzo.

[1] *Settepergole*—Seven Pergolas—the name of several farms in Wälsch-Tirol. *Pergola* is the name for a vine trellised to form an arbour, all over Italy.

[2] This effect has often been noticed here by travellers.

steeple, of which he never saw the like; the folk he met by the way were all strangers, and stared at him as at one who comes from far. He wandered up and down all the day, and everything was yet strange. At evening the men came back from the fields, and again they gazed at him estranged: once he made bold to ask them for 'Zansusa,' the companion of his boyhood, but they shrugged their shoulders with a 'Chè Zansusa?' and passed on. He asked again for 'Piero,' almost as dear a friend, and they pointed to a 'Piero' with not one feature like *his* Peter. Once again he asked for 'Franceschi,' and they pointed to a grave, where his name was written indeed—'Franceschi,' who but the day before had walked with him in full life and health, to hang a fresh wreath on Filomena's cross! Ah! there was Filomena's cross, but how changed was that too! the bright gilding, on which his savings had been so willingly lavished, was tarnished and weather-worn, and not a leaf of his garland remained round it. He wandered no further, nor sought to fathom the mystery more; he knelt on the only spot of earth that had any charm for him. As his knees touched the hallowed soil consoling thoughts of her undying affection overflowed him. 'Here we are united again,' he said; 'in a little while we shall be united for ever.' 'At last have I found thee! these fifty years I have sought thee in vain!' The moonbeam kissed his forehead as he looked up, and the moonbeam bore her who had spoken. A fair form she wore, but still it was not the form of

Filomena. 'Who are you, and wherefore sought you me?' he asked. 'I am DEATH,' replied the pale maiden, 'and for fifty years I have sought thee to lead thee to Filomena.' She beckoned as she spoke, and willingly he followed her whither the moonbeam led.

The village of Caldonazzo, with its ancient castle, is another ornament of the lake. Further south is the village of Lavarone, or Lafraun, accessible only to the pedestrian. A house close to the edge of a little lake here is pointed out, which in olden time was the residence of two brothers, the owners of the meadow over which the lake is now spread. These two could never agree; their strife grew from day to day, till at last one night they called each other out to settle their quarrels once for all by mortal combat. The noise of the strife within had made them oblivious to the strife of the elements which was waging without. The gust which entered as the eldest turned to open the cottage door, and the blinding rain, drove them back; even their fierce passions seemed mastered by the fiercer fury without. In silence they returned into the room, and neither cared to raise his voice amid the angry voices of the storm, which now made themselves heard solemnly indeed. In sullen silence they passed the night, and during the silence there was time for reflection; each would have been glad to have backed out of the promised fight, but neither had the courage to propose a reconciliation. Sullenly they rose with the morning light; the pale gold rays rested on the trees, now calm and tranquil, and both shuddered to carry their vengeance out on to the fair scene; but neither dared speak, and once more the eldest opened the door. This time it was not the rain descending from above which drove him back; it was the flood rising from beneath! The Centa torrent had overflowed. The disputed meadow had become a lake, and with their united efforts they scarcely kept the waters banked out. The community of labour, of danger, and of distress, ended the strife; and though their worldly possessions were lost to them for ever, they had found a greater boon, the bond of fraternal charity.

I must pass over Levico, near which the Brenta has its source, and the intervening villages; but Borgo di Val Sugan' demands our attention for its beautiful situation. The view over both may be enjoyed by mountain climbers from the neighbouring height of Vezzena. Borgo is commonly called the Italian Meran, for its likeness with that favourite watering-place. Its buildings extend over both sides of the Brenta, being united by a massive stone bridge, built in 1498. Those on the left bank were nearly destroyed by fire in 1862, but the rebuilding has been carried on with great spirit. Its ecclesiastical buildings do not date far back; the rebuilding of the parish church in 1727 nearly obliterated all traces of the earlier edifice; its chief glories are three paintings it possesses, one by Titian's brother, one by Karl Loth, and one by Rothmayr. The fine campanile was added in 1760. There is also a Franciscan convent, but it does not date back further than 1603, there is the following curious tradition of its origin.

The Sellathal leading to Sette Comuni, is narrowed by two mighty cliffs—the Rochetta on the south, and the Grolina on the north, adorned with the ruined Castel San Pietro,[1] seemingly perched above all human reach. On a green knoll beneath it stand the lordly remains of Castel Telvana; its frescoes are now nearly faded away, only a room here and there is habitable; but its enduring walls and towers show of what strength it was in the days long gone by—days such as those in which Anna, wife of Siccone di Caldonazzo, defended it with so much spirit against all the might of Friedrich *mit der leeren Tasche,* that she obtained the right to an honourable capitulation. It was bought by the Counts of Welsburg in 1465, and henceforth it became an abode of pleasure rather than a mere fortress. Count Sigmund von Welsburg, who was its master towards the end of the sixteenth century, was particularly disposed to make his residence in their midst a boon to the inhabitants of Borgo, and entered heartily into all the pastimes of the people. It happened thus that the Carneval procession of the year 1598 was invited to take the Castel Telvana for its bourne; and that the women might not be fatigued by the ascent, the Count gallantly provided them all with horses from his own stud. The valley resounded with merriment as they wended their way up in their varied and fantastic

[1] Two bronze statuettes of Apollo were found here in June 1869.

attire. Arrived at the castle, good cheer was provided, which none were slow to turn to account, and the return was commenced in no less boisterous humour. At the most precarious spot of the giddy declivity, the courage of the foremost rider forsook her; the Count's high-couraged charger, which she bestrode, perceiving the slackened pressure on the rein, grew nervous and bewildered too, and uneasy to find himself for the first time subjected to devious guidance. The indecision of the first fair cavalier alarmed her sister, who followed next behind—a shriek was the expression of the alarm, which communicated itself to the next rider, and in a moment a panic had possessed the whole calvacade, or nearly the whole; for the few who here and there still retained their presence of mind were powerless to make those before them advance, or to keep back the threatening tramp of those behind. The Count saw the danger, and the one remedy. First registering a vow, that if he succeeded in his daring enterprise he would build a convent to the honour of God and St. Francis, he set out along the brink of the narrow track, where there was scarce a foot-breadth between him and the abyss, past the whole file of the snorting horses and their terrified burdens. He had this in his favour, that every denizen of his stable recognised him as he went by, and his presence soothed their chafing. Arrived at last safely at the head of the leading steed, his hand on its mane was enough to restore its confidence; securely he led it to the full end of the dangerous pass, and all the others followed in docile order behind. The Count did not forget his vow, nor would he in his gratitude allow any other hand to diminish the outlay he had undertaken. The convent buildings are now in part turned to secular uses, though part is also used for a hospital, where all the sick of the town are freely tended. In the church is an altar-piece of Lazarus begging at the gate of Dives, by Lorenzo Fiorentini, a native artist.

The pass I have mentioned between the Rochetta and the Grolina—the importance of which as a defence was not unknown to the Romans, of whose remains the town possesses a considerable collection dug up at different times—was not without its share of work in the French invasions of 1796 and 1809. In the former, a handful of Tirolese successfully repulsed five hundred of the enemy in an obstinate encounter of three hours' duration. In the latter, the place was attacked by tenfold greater numbers. General Ruska was so infuriated, not only by their determined and galling fire, but by the derisive shouts and gestures of the mountaineers, who carried their daring so far as to fling the dead bodies of the soldiers they had killed down under the wheels of his carriage, that he ordered the pillage and destruction of the town. His guns were ready planted to pour out their murderous fire, when the parish priest, heading a procession of aged housefathers, came to implore him to spare their homes. At the same moment news was brought him that two Austrian battalions were advancing with dangerous haste. One or other of the considerations thus urged effected the deliverance of the town, which was only required to buy itself off at the price of a large supply of provisions.

Borgo has further advantage of the mineral spring of Zaberle, and a creditable theatre. Silk-spinning is again the chief industry of the place; and there are several so-called *Filatoriums*, employing a great number of hands. The most remarkable excursions in the neighbourhood are to the deserted hermitage of San Lorenzo and the stalactite caves of Costalta, both in the Sellathal, whence there is a path leading to the curiously primitive and typically upright community of the *Sette Comuni*.

Pursuing the valley further in its easterly course, I must not omit to mention Castelalto, not only remarkable for its share in the mediæval history of Tirol, but for being still well kept up. At Strigno, one of the largest hamlets of the valley, is another ancient castle, which after its abandonment in the fourteenth century acquired the name of Castelrotto. The parish church, rebuilt in 1827, contains a Madonna del Rosario by Domenichino; and a Mater Dolorosa in Carrara marble, by the Venetian sculptor Melchiori. This is the generally adopted starting-place for the Cima d' Asta, the highest peak of the Trentino (8,561 feet), and commanding a panorama of exceptional magnificence. Under favourable circumstances it is reached within thirty hours, sleeping in the open at Quarazza. The interest of the way is heightened by two considerable lakes; the lower, that of Quarazza, closed in by wall-like cliffs, is fed by a cascade from the higher lake, which receives several torrents. Near the summit is a garnet quarry. Just below Strigno is another inhabited castle, that of Ivano, belonging to the Count of Wolkenstein-Trostburg, who makes it a summer residence. The church is dedicated to S. Vindemian; near it was once a hermitage. Further down the valley is Ospedaletto, famous in border warfare, and once a hospice for travellers, served by monks, still a mountain-inn with a chapel attached. Grigno has another once-important castle. S. Udalric, Bishop of Augsburg, had occasion to pass through the village on his way to Rome in the time of Pope Sergius III. (A.D. 904-11), and left behind him so profound an impression of his sanctity, that the devotion of the people to his memory has never diminished. In the eleventh century a chapel was built in his honour, with the picturesque instinct of the people of that date, on the steep way leading to Castel Tesino. It was always kept in good condition till 1809, when it was desecrated by the French soldiery. It was restored within ten years, and a rustic piazza in front planted with lime trees, which have at the present time attained considerable dimensions. In July 1869, processions consisting of more than four thousand villagers met at this shrine, to pray for deliverance from the heavy rains, which were causing the inundation of their homesteads.

From Grigno there is a path which few persons however will be tempted to follow, across the so-called Canal San Bovo, to Primiero, a country which has already been so ably laid open to the tourist that I need not attempt a fresh description of its beauties. If any one penetrates its recesses as far as the village of Canal San Bovo, I think they will not be sorry to have been advised to ask for a certain Virginia Loss, who has a touching story to tell them of her adventures. On a stormy day, the last of October 1869, she was making her way, though only thirteen, with her mother and another woman, along the dangerous path leading hither from the Fleimserthal, following their occupation of carriers. They had passed Panchià and Ziano, and were in the midst of the verdant tract known as the Sadole. The fierce wind that blew exhausted her poor mother's strength, and she saw no help but to lay down her burden by the way, and try to reach home with bare life. Domenica Orsingher, the other woman, however, who had already got on a good way beyond her, no sooner learned what she had done than, considering what a loss it must be to her, with a humble heroism went back to fetch the pack intending to carry it in addition to her own! The next day some men travelling by the same path found her body extended by the wayside. She had died of cold and exhaustion.

> The land is strong with such as these,
> Her heroes' destined mothers.

Further along they found Elisabetta Loss and her daughter huddled together. On carrying the bodies to

Cauria they succeeded in reviving only the child. Virginia has a tragic story to tell of; of how her mother sank to her rest, and her own unavailing and inexperienced efforts to call her to life; then the horror of the approaching night, the snow storm in which she expected to be covered up and lost to sight, yet had not strength to move away; and, worst of all, the circling flight of crows and ravens which she spent her last energies in driving with her handkerchief from her mother's face; and yet the presence of death, solitude and helplessness, made the approach of even those rapacious and ill-omened companions seem almost less unwelcome. The insensibility which ensued was probably the most welcome visitant of all.

Le Tezze is a smaller village than Grigno, but one that has done good service to the patriotic cause, having many a time stayed the advance of invading hosts; and never more successfully than in the latest Garibaldian attempt on the Trentino, upon the cession of Venice by Austria after Sadowa. The tombs of the bold mountaineers who fell while driving back the tenfold numbers opposed to them are to be seen appropriately ranged along the stony declivity they defended so well. These graves are yearly visited by their brethren on the 14th of August.

> They fell devoted and undying.
> The very gale their deeds seems sighing;
> The waters murmur forth their name,
> The woods are peopled with their fame,
> The silent pillar, lone and gray,
> Claims kindred with their sacred clay.

Le Tezze is the last Tirolean village of the valley, and the seat of the Austrian custom-house against Italy. On the other side of this frontier is the interesting Italian town of Primolano, whence there is an easier way into Primiero-thal than by crossing the Canal San Bovo. Val Sugana retains more of the German element than any other district of Wälsch-Tirol.

Judicarien or Giudicaria bifurcates westwards and south-westwards from the Etschthal opposite Val Sugana. Its first (south-west) division is called the Sarcathal and reaches to the Lago di Garda. Though no part of the beautiful Italian lake actually belongs to Tirol the town of Riva overlooks it; the country round is most productive in wine, silk, lemons, figs, and other fruits. Its pleasant climate, the warmest in all Tirol, is due not only to its southern latitude, but also to its being the lowest land of the principality. Innsbruck is 1,820 feet above the sea-level, Riva but 220. From the western division of Giudicaria there branch out northwards Val Rendena, north-westwards Val Breguzzo and Val Daone, and southwards Val Bona. The Val di Ledro or Lederthal, forms a parallel return towards the Garda-See. Here an attempt at invasion headed by Garibaldi was repulsed by the Innsbruck Student-brigade in 1866 at a pass called Bezzecca.

Giudicaria is little explored yet it contains some choice scenery and traditions. Castel Madruzz, which can be visited from Trent, is one of its most ancient and important castles. From the twelfth to the seventeenth century, the family which inhabited it and bore its name takes a foremost place in Tirol's history. In the church are shown the portraits of seven of the family ascribed to Titian. From 1530 to 1658 four of its members occupied the See of Trent, and were successively invested with the Cardinalitial dignity. Cardinal Karl Madruzz became the last of his house. All his kindred having died without heirs, he applied to Rome for permission to marry—a dispensation which we have seen once before accorded in favour of a Tirolese prince. Cardinal Madruzz preferred his suit successively before Urban VIII., Innocent X., and Alexander VII., and at last obtained it, coupled with the proviso that he should only marry in his own station. As this did not accord with his intentions, the favour so tardily granted was never acted on. This fine castle had fallen into sad neglect but it is being restored. From its deserted terraces a glorious view is obtained, which takes in the two lakes of Toblino to the north, and Cavedine to the south, both being fed by the same torrents. Round the Lago di Cavedine lie the flowery slopes which bear the name of Abraham's Garden. The Lake of Toblino is broken into by a picturesque promontory, bearing the castellated villa of the Prince-Bishops of Trent; though on flat ground, the round turrets at the angles with their pointed caps afford a wonderful relief to the landscape. The village is called Sta. Massenza, from the mother of S. Vigilius, who died here in the odour of sanctity, 381. Her relics were translated to Trent, 1120. At the foot of the height on which stands Schloss Madruzz is a double chapel, on the model of the Holy House of Loreto, the legend being inscribed on the walls.

At the westernmost reach of Giudicaria, the Rendenathal branches off towards Val di Sole. It was the cradle of the evangelization of Tirol, for here S. Vigilius suffered martyrdom, 405, and the valley is rife with traditions of him. He appears to have been stirred with zeal for the propagation of the faith at a very early age; and his piety and earnestness were so apparent that he was consecrated Bishop of Trent at the age of twenty. He made many conversions, and built a church to SS. Gervasius and Protasius, A.D. 375. But he was not content with establishing the faith here, and sending out missionaries hence; he would wander himself on foot through all the valleys where paganism still lurked, overturning idols and building Christian sanctuaries—more than thirty trace their origin to his work. Nowhere did he meet with so much opposition as in the Rendenathal, which was the last to accept the yoke of Christ. But he was untiring in his apostolic labours, nor could he rest while *one* token of a false religion remained erect. It is not to be supposed that, though he made many fervent converts, he effected all this without also exciting the opposition and fury of those whose teaching he had come to supersede. Yet though many were the snares set for him, no conspiracy against him succeeded till he had cast down the last idol. It was at Mortaso, one of the remotest villages of this secluded dell, he stood announcing the 'glad tidings' of the Gospel from the pedestal of the image he had overthrown, and the population crowded round, earnestly garnering in his words. He had left off preaching, and just raised his hands in benediction, when a body of heathen men and women, who had long determined to compass his end, rushed upon the scene from the surrounding grove, and stoned him with the fragments of the image he had overthrown. His hearers would have defended him, but he knew that his hour was come, for his work was accomplished; and forbidding all strife, he knelt down, and folding his arms on his breast meekly rendered up his spirit, while his constancy won many to the faith. His disciples reverently gathered his remains and bore them to Trent; but as soon as his murderers were aware of their intent, they set out to follow them. The Christian party, delayed by the weight of their burden, found that their pursuers were fast gaining ground. In this strait, says the legend, they called upon the rocky wall before them—

> Apritevi, O sassa,
> Che S. Vigilio passa,

and behold before them suddenly appeared a cleft in the rock, through which they passed in safety, and which is pointed out to this day. Another narrow cleft is pointed out near Cadine, which is said to have been rent asunder at his bidding, when once, at an earlier stage of his labours, he deemed it right to flee from those who would have taken his life. The *Acqua della Vela* now passes through it, and a dent is shown which is said to mark the place where the saint impressed his hand on the obedient stone. It was this suggested to the bearers of the bier to make a similar appeal on behalf of his relics. It is commonly reported that in Mortaso the bread never rises properly; and they couple with it this tradition, that when the pieces of the broken idol sufficed not for all who would attack the saint, the women brought out loaves from the oven to complete the work.

The Rendenathal also preserves the memory of S. Julian, called also Sent Ugiano and San Zulian in local dialect. His legend says he lived with his parents in an outlying house. On one occasion, at the time of day when they were usually at work in the fields, he heard the sound of persons entering the house, and turned and slew them, and only found afterwards that it was his parents whose lives he had taken.[1] Struck with horror he devoted himself to a life of penance, and made a vow to live so far from the habitations of men that he should no more hear the cheerful crowing

[1] Very like and very unlike the legend of S. Giuliano I met in Rome (*Folklore of Rome*), where he was supposed to be a native of Albano, and to have passed his penitential time at Compostella. G. Schott, Wallächische Märchen, pp. 281 and 375, gives a similar legend applied to Elias in place of St. Julian.

of the cock or the holy chime of the church bells. After his death the people found that angels had planted roses on his grave which bloomed in winter, and they observed that no venomous reptile ever rested on it, while earth taken from it cured their sting. So they built a chapel in his honour on the border of the little lake which bears his name, at the opening of Val Génova. Another interesting church in the same locality is that of Caresolo. Its exterior walls are adorned with frescoes bearing date 1519, and inside is an inscription recording that it was restored by the munificence of Charles Quint. At Pelugo, near Tione, where the Rendenathal branches off, he found the castle in possession of a Jew, and so indignant was he to find a once Christian fortress so occupied, that he had him immediately ejected and the place exorcised. Here, as also at Massimeno and Caderzone, all inconsiderable mountain villages, new churches were consecrated during the Bishop of Trent's visitation in August 1869, showing that the spirit of S. Vigilius had not died out. In the Pfarrkirche at Condino is a *Muttergottesbild*, presented in 1620 by a parishioner who averred he had seen it shed tears. Of the church of Campiglio the legend runs, that when it was building, the people being much distressed by a dearth, and their means hardly sufficing, the angels used to bring stone, wood, and other materials in the night; and one pillar is pointed out which was raised before the eyes of the builders in broad day by invisible hands. The inn here occupies a hospice built by the Templars, hence its imposing appearance. Colini, who was locally called the Hofer of Wälsch-Tirol, for his brave leadership of his countrymen in 'the year nine,' kept it till his death in 1862. At Pinzolo is a thriving glass-house, supported by Milanese capital and Venetian art and industry.

Riva, at the head of the Garda-See, is one of the most charming spots in Tirol. Its German name of *Reif* is not a mere corruption of the Italian name; it is an old German word, having the same signification, of a shore. The parish church is a really handsome edifice, and a great ornament to the town and neighbourhood. Outside the town is a curious octagonal church of the Immaculate Conception, built to enclose a wonder-working picture of the Blessed Virgin, by Cardinal Karl von Madruzz, who also founded a House of Friars Minor to attend to the spiritual necessities of the many pilgrims who came to visit it. The churches of S. Roch and S. Sebastian were built on occasion of visitations of the plague in 1522 and 1633. The neighbourhood supplies the whole of Tirol with twigs of olive to use in the office of Palm Sunday, and all kinds of southern produce grow on the banks of the lake. It was long considered the highest latitude at which the olive-tree would grow, but it has since been successfully cultivated as far north as Botzen. In order to gain a full enjoyment of the beautiful scenery around, the Altissimo di Nago should be ascended by all who have the courage for a six or seven hours' climb. From San Giacomo, however, where there is a poor *Wirthshaus* and chapel, reached in not more than two hours, the scene at sunrise is one of inconceivable beauty. Behind are ranges beyond ranges and peaks beyond peaks of lordly alps. Before you lies the blue Lago di Garda, and the vast Lombard plains studded with fair cities, amid which you will not fail to distinguish Milan, which some optical illusion brings so near that it seems it would take but an easy morning's walk to reach it.

On the way hence to Mori, at about half distance, lies Brentonico, with a new church perched picturesquely as a mediæval one on a bold scarped rock. The old parish church has a fine crypt. The Castello del Dosso Maggiore is a noble ruin. There is a bridge over a deep defile in the outskirts, called the *Ponte delle strege*—the Witches' Bridge—being deemed too daring for human builders. Mori, though named from its mulberry trees, is more famed for its tobacco, which is reckoned the best grown in Tirol.

Wälsch-Tirol has many traditions, customs and sayings, which differ from those of the rest of the Principality, more resembling those of Italy, and some of which it cannot be fanciful to trace back to an Etruscan connection. Some bear the impress of the Roman occupation, and all are strung together by an overpowering Germanic influence.

The most prominent group—and their special home,

I am assured, clusters round the Dolomite mountains—are those concerning certain beings called 'Salvan' and 'Gannes;' and traditions about 'Orco.' A local collector of such lore, to whom I am chiefly indebted for the above fact, is inclined to identify the 'Salvan' with 'Orco;' but I think it can be shown that they are distinct ideas. Both are only ordinarily, not always malicious, but the 'Salvan' is one of a number of sprites, Orco has the dignity of being one by himself. The Salvan in some respects takes the place of the wild man of the North, and of the satyr whom I also found called in Rome 'salvatico' and 'selvaggio.'[1] 'Orco' clearly takes the place of Orcus in Italy; and that of the 'Teufel' in German legend. Yet so are the traditions of neighbouring peoples intermingled, that the Germans, not content with their own devil, have sprightly imitations of Orco in their 'Nork' and Lorg, softened in the intermediate Deutsch Tirol into Norg.[2] In Norway the same appellation is found, hardened into Nök, Neck, Nikr,[3] which seems to bring us round to our own 'Old Nick;' for in Iceland he is 'Knikur,' and, perhaps, he gave his name to Orkney.[4]

---

[1] *Folklore of Rome*, p. 320.

[2] I need not repeat the characteristics of the Tirolean Norg, which I have given in the translation of the 'Rose-garden' in *Household Stories from the Land of Hofer*.

[3] Thorp's *Northern Mythology*, vol. ii. pp. 20–2.

[4] Though of course mere similarity of sound may lead one absurdly astray; as if any one were to say that the old fables of rubbing a ring to produce the 'Slave of the ring' was the origin of the modern substitute of *ringing* to summon a servant!

It is curious, in tracing the seemingly undoubtable connection between the Norg and Orco, to observe that though the Norg possesses almost invincible strength, and often prevails against giants, yet in stature he is always a dwarf, while Orco himself is considered a giant. But then it is the one essential characteristic of Orco which forms the link between all conceptions of him, whether men call him Orco, Nork, or Nyk, that he is a deceiver ever; a liar from the beginning; whenever he appears it is continually under some ever-changing, not-to-be-expected form, and only the wise guess what he is before it is too late.[1] Thus it happened to two young lads of Mori, who had been up the mountains to visit their sweet-hearts, and coming back, they met Orco prowling about after his manner when all good people are safe in bed asleep—this time in the form of an ass. The Mori lads, never thinking but that it was a common ass, jumped on its back. They soon found out their mistake, for Orco quickly resented their want of discrimination, and cantered off with them past an old building which had once been a prison, and skilfully chucked them both in at the window. It was some days before they contrived to crawl out again, and not till they were nearly starved.

---

[1] Again, Mr. Cox (*Mythology of the Aryan Nations*, vol. ii. p. 221 et seq.) says, 'the Maruts or storm winds who attend on Indra . . became the fearful Ogres in the traditions of Northern Europe . . they are the children of Rudra, worshipped as the destroyer and reproducer and . . like Hermes, as the robber, the cheat, the deceiver, the master thief.'

But we have in English another affinity with 'Orco,' besides 'Old Nick;' we have seen him take the place of our 'ogre' in deed as well as in name in the Roman fairy tales, and in Italy he is also the bugbear of the nursery which we have almost literally in 'Old Bogey.' And now Mr. I. Taylor has found another affinity for him if he be justified in identifying our 'ogre' with "the Tatar word, '*ugry*,' a thief." [1]

To return to Orco's place in Tirol, we find his name assumes nearly as many transliterations as his external appearance assumes changes. In Vorarlberg they have a Dorgi or Doggi (i being the frequent local abbreviation for the diminutive *lein*,—*klein*), there considered as one personation of the devil. The Doggi spreads over part of Switzerland, and overflows into Alsace as the Doggele.[2] In the zone of Tirol where the Italian and German elements of the population mingle, there is a class of mischievous irrepressible elfs called Orgen; soft, and round, and small, like cats without head or feet, who establish themselves in any part of a house performing all sorts of annoyances, but who are as afraid of egg-shells as the Norgs in other parts are said to be. Their chief home is in the Martelthal, south of Schlanders in the Vintschgan, and their name is devoted to the brightly shining peak seen from it— the Orgelspitz. In the Passeyer, on the north side of the Vintschgan, they go by the name of Oerkelen.

[1] *Etruscan Researches*. p. 376 and note.
[2] Stöber, *Sagen des Elsasses*, p. 30.

Since we have seen him, too, divested of his 'r' in Doggi from Vorarlberg to Alsace, and the Germans have already given him an L in Lorg, he assumes a mysterious likeness to Loki himself, and as a sample of how elastic is language, and how misleading are mere sounds, though for no other purpose, it might be said, we had found in this Doggi a relation of the dog who guards the entrance to the regions of Orcus!

The Salvan and Gannes, as described by the local observer above alluded to, seem to partake very much of the character of the good and evil genii of the Etruscans, though the traditions that remain of them refer almost exclusively to their action on this side the grave. 'Their Etruscan appellation,' says Mr. Dennis, 'is not yet discovered;'[1] when it is, it will be very satisfactory if it has any analogy with 'Gannes.'[2] The Gannes were gentle, beauteous, beneficent beings, delighting in being helpful to those they took under their protection; harmful to none. The Salvans were hideous, wild, and fierce, delighting in mischief and destruction, with fiery serpents for their chief companions. They seem to have done all the mischief they could as long as their sway lasted, but they were scared by advancing civilization; and I have a ludicrous

---

[1] *Cities of Etrusca*, vol. ii. p. 65-8.

[2] Selvan, at all events, is a word which, Mr. Isaac Taylor observes, is of frequent occurrence in Etruscan inscriptions (*Et. Res.* pp. 394-5), and its signification has not yet been fixed. And may not *Gannes* have some relation with Kan or Khan (p. 322)?

description of how they stood gazing down in stupid wonderment from their Dolomite peaks, when the first ploughs were brought into use in the valleys.

Schneller, who with all his appreciation of Wälsch-Tirol, looks at its traditions too much through German spectacles, gives us some little account of these beings too.[1]

He has also a 'Salvanel,' who seems a male counterpart of his Gannes, helpful and soft-natured, with no vice save a tendency to steal milk. In return he teaches mankind to make butter and cheese, and other useful arts, and is specially kind to little children; his name bears some relation with the local word for the

---

[1] It is very disappointing that he has translated the great bulk of his vast collection of fiabe ('*fiaba*' in North Italian answers to the 'favola' of Rome) so utterly into German that, though we find all our old friends among them, all the distinctive expressions are translated away, and they are rendered valueless for all but mere childish amusement. Thus it is interesting to find in Wälsch-Tirol a diabolical counterpart of the Roman story of 'Pret' Olivo,' but it would have rendered it infinitely more interesting had the collector told us what was the word which he translates by 'Teufel,' for it is the rarest thing in the world for an Italian to bring the personified 'Diavolo' or 'Demonio' into any light story. In the same way it is interesting to find all the other tales with which we are familiar turn up here, but the real use of printing them at length would have been to point out their characteristics. What was the Italian used for the words rendered in the German by 'Witch?' Was it 'Gannes' or 'strega?' or for 'Giant' and 'Wild man:' was it 'l'om salvadegh' or 'salvan' or 'orco?' I cannot think it was 'gigante.' But all is left to conjecture. Among the few bits of Italian he does give are two or three 'tags' to stories, among them the one I met so continually in Rome 'Larga la foglia'—(it was still 'foglia' and not 'voglia') word for word.

'Jack-'o-lantern' reflection from glass or water. But he found also the 'Salvan' in his pernicious character under the names of 'Bedelmon,' 'Bildermon,' and 'Salvadegh.' But the most pernicious spirit that came in his way was the 'Beatrik,' who is an unmitigated fury,[1] and the natural enemy and antagonist of a gentle, helpful, beauteous spirit called Angane, Eguane, and Enguane, but possessed with his German ideas, he saw in the being so designated nothing but 'a witch, or perhaps a fairy-natured being.'[2] In another page he pairs them off more fairly with the 'Säligen Fräulein' of Germany. Here is a story of their ways which was given me, but I do not know if it was founded on his at page 215, or independently collected:—A young woodman was surprised one day to meet, in the midst of his lonely toil, a beautiful maiden, who nodded to him familiarly, and bid him 'good day' with more than common interest. Nor did her conversation end with 'good day;' she found enough to prattle about till night fell; and then, though the young woodman had been sitting by her side instead of attending to his work, he found he had a bigger faggot to carry home than he had ever made up with all his day's labour before. 'That was a sweet maiden, indeed,' he mused on his way home. 'And yet I doubt if she is all right. But her talk showed

[1] Dr. Steub, in his *Herbsttagen in Tirol*, shows that the Beatrick may be identified with Dietrich von Bern.

[2] Though nothing would seem simpler than to suppose the word derived from the Euganean inhabitants who left their name to Val Sugana.

she was of the right stuff to make a housewife; but then Maddalena, what will she say? ha! let her say what she will, she won't stand comparing with her! I wonder if I shall see her again! And yet I don't think she's altogether right, either.' So he mused all through the lonely evening, and all through the sleepless night; and his first thought in the morning was of whether he should meet that strange maiden again in the wood. In the wood he did meet her, and again she wiled away the day with her prattle; and again and again they met. Maddalena sat at home weeping over her spinning-wheel, and wondering why he came no more to take her for a walk; but Maddalena was forgotten, and one day it was her fate to see her former lover and the strange maiden married in the parish church. The woodman was not surprised to find his seiren the model of a wife. The house was swept so clean, the clothes so neatly mended, the butter so quickly churned, that though all the villagers had been shy of the strange maiden, none could deny her excellent capacity. The woodman was very well satisfied with his choice; but as he had always a misgiving that there was something not quite right with her, he could not help nervously watching every little peculiarity. It was thus he came to notice that it was occasionally her custom to lay her long wavy tresses carefully outside the bedclothes at night; he thought this odd, and determined to watch her. One night, when she thought him asleep, and he was only feigning, he observed that she took a little box of salve from under her pillow, and rubbing it into her hair, said, *Schiva boschi e schiva selva* (shun woods and forests), and then was off and away in a trice. Determined to follow her, he took out the box of salve, and rubbing it into his hair, tried to repeat her saying, but he did not recall it precisely, and said instead, *Passa boschi e passa selvi* (away through woods and forests), and away he went, faster than he liked, while his clothes and his skin were torn by the branches of the trees. He came, however, to the precincts of a great palace, where was a fresh green meadow, on which were a number of kine grazing, and some were sleek and well-favoured, while some were piteously lean; and yet they all fed on the same pasture. The palace had so many windows that it took him a long while to count them, and when he had counted them he found there were three hundred and sixty-five. He climbed up and looked in at one of them—it was the window of a great hall, where a number of *Enguane* were dancing, and his wife in their midst. When he saw her, he called out to her; but when she heard his voice, instead of coming she took to flight, nor could he overtake her with all his strength for running. At last, after pursuing her for three days, he came to the hut of a holy hermit, who asked him wherefore he ran so fast; and when he had told him, the hermit bid him give up the chase, for an Enguane was not a proper wife for a Christian man. Then the woodman asked him to let him become a hermit too, and pass the remainder of his life under his guidance. To this the hermit consented; so he built him a house, and they lived together in holy contemplation. One day the woodman told the hermit of what he had seen when he went forth to seek his wife; and the hermit told him that the palace with three hundred and sixty-five windows represented this temporal world, with its years of three hundred and sixty-five days; but the fresh green meadow was the Church, in which the Redeemer gave His Flesh for the food of all alike; but that while some pastured on it to the gain of their eternal salvation, who were represented by the well-favoured kine, there were also the perverse and sinful, who eat to their own condemnation, and were represented by the lean and distressed kine.[1]

It is less easy to collect local traditions in Wälsch-Tirol than in any other part of the principality, but legends and marvellous stories exist in abundance; and so long as the institution of the *Filò* (or out-house room where village gossips meet to spend their evenings

---

[1] It is curious to observe the story pass through all the stages of the supernatural agency traditional in the locality; first the good genius of the Etruscans merging next into the Germanic woodsprite, then assuming the vulgar characteristics of later imaginings about witchcraft, and then the Christian teaching 'making use of it,' as Professor de Gubernatis says, 'for its own moral end.'

in silk-spinning and recounting tales) last, they will not be allowed to die out:[1] it is said that there are some old ladies who can go on retailing stories *by the week together!* And though by the nature of the case these gatherings must consist almost exclusively of women, yet it is thought uncanny not to have any man about the place; in fact, that in such a case *Froberte*[2] is sure to play them some trick. They narrate that once when this happened, one of the women exclaimed, 'Only see! we have no man at all among us: let's be off, or something will happen!' All rose to make their escape at the warning, but before they had time to leave, a *donna Berta* knocked and came in. 'Padrona! donna Berta dal nas longh,'[3] said all the women together, trying to propitiate her by politeness; and the nearest offered her a chair. 'Wait a little, and you'll see another with a longer nose than I,' re-

---

[1] A collection of the 'Costumi' of Tuscany I have, without a title-page, but I think published about 1835, laments the growing disuse in Lunigiana (*i.e.* the country round the Gulf of Spezia, so called from Luna, an Etruscan city, but ' not one of the twelve,' and including Carrara, Lucca, and Pisa) of the practice of recounting popular traditions at the *Focarelli* there. These seem to be autumn evening gatherings round a fire, but in the open air, often on a threshing-floor; while the able-bodied population is engaged in the preparation of flax, and some are spinning, the boys and girls dance, wrestle, and play games, and the old crones gossip; but now, says the writer, they begin to occupy themselves only with scandalous and idle reports, instead of old-world lore.

[2] My readers will perhaps not recognize at first sight that this is a corruption of *Frau Bertha*, the *Perchtl* whom we met in North Tirol. In the Italian dialects of the Trentino she is also called *la brava Berta* and *la donna Berta*.

[3] 'Your servant! Mistress Bertha of the long nose.' Such was supposed to be the correct form of addressing the sprite.

plied Froberte; and as she spoke, a second *donna Berta* knocked and entered, to whom the women gave the same greeting. 'Wait a bit, and you'll see another with a longer nose than I,' said the second *donna Berta*; and so it went on till there were twelve of them. Then the first said, 'What shall we be at?' To which the second made answer, 'Suppose we do a bit of washing:' and the others agreeing, they told the women to give them pails to fetch water with; but the women, knowing that their intention was to have suffocated them all in the wash-tubs, gave them baskets instead. Not noticing the trick, they went down to the Etsch with the baskets to fetch water, and when they found that all their labour was in vain, they ran back in a great fury; but in the meantime the women had all escaped to their home, and every one was safe in bed with her husband. But a Forberte came to the window of each and cried, 'It is well for you you have taken refuge with your husband!' The next night the women were determined to pay off the *brava Berta* for the fright they had had, so they got one of their husbands to hide himself in the crib of the oxen; had he sat down with them, the Froberte would not have come at all. Not seeing him, Froberte knocked and came in, and they greeted her and gave her a chair, just as on the previous night: and the whole twelve soon arrived. Before they could begin their washing operations, however, the man sprang out of the crib, and put them to flight with many hard blows; so that they did not return for many a long day. The last day of Carneval was called *il giorno delle Froberte*, probably because many wild pranks in which sober people allow themselves to indulge on that day of licence were laid on the shoulders of Mistress Bertha. But it is also said, that since the sitting of the Holy Council of Trent, the power for mischief of these elves has grown quite insignificant. Here are some few specimens of the multifarious stories of the *Filò*.[1] Once there was a man and his wife who had two daughters: one pretty, but vain and malicious; the other ugly, but docile and pious. The mother made a favourite of the pretty daughter, but set the ugly one to do all the work of the house; and though she worked from morning to night, was never satisfied with her. One day she sent her down to the stream to do the washing; but the stream was swollen with the heavy rains, and had become so rapid that it carried off her sister's shift. Not daring to go home without it, she ran by the side of the stream, trying to fetch it back. All her pains were vain; the stream went on tumbling and roaring till it swelled out into a big river, and she could no longer even distinguish the shift from the white foam on which it was borne along. At last, hungry and weary, she descried a house, where she

---

[1] Many of these concern the earthly wanderings of Christ and his apostles. I have given one of the most sprightly and characteristic of Schneller's, too long to be inserted here, in *The Month* for September, 1870, entitled 'The Lettuce-leaf Barque.'

knocked with a trembling hand, and begged for shelter. The good woman come to the door, but advised her not to venture in, for the *Salvan* would soon be home; but the child knew nothing about the *Salvan*, but a great deal about the storm, and as one was brooding, and night coming on, she crept in. She had not been long inside, when the *Salvan* came home, also seeking shelter from the storm. 'What stink is this I smell of Christian flesh?' he roared; and the child was too truthful to remained concealed, and so came forward and told all her tale. The *Salvan* was won by her artlessness, and not only allowed her a bed and a supper, but gave her a basketful of as much fine linen as she could carry, to make up for her loss. When her pretty sister saw what a quantity of fine linen the *Salvan* had given her, she determined to go and beg for some too; but when the *Salvan* saw her coming, he holloaed out, 'So you're the child who behaves so ill to your sister!' and he gave her such a rude drubbing, that she went back with very few clothes on that were not in rags.

In selecting a specimen or two of the *fiabe* I will take first a group going by the name of 'Zuam' or 'Gian dall' Orso' (Bear-Johnny),[1] because the Wolf-boy group is a very curious one, and this is our nearest

---

[1] Gathered for the above-named collection by Herr Zacchea of the Fassathal, in the Val di Non, Lederthal, and Val Arsa.

approach to it,[1] though it deals with a bear-child and not a wolf-child;[2] and because we have already found Orso and Orco confounded in Italian folk-lore at Rome. The following is from Val di Non :—A labourer and his wife had their little boy out with them as they worked in the fields. A she-bear came out of the woods and carried him off. She treated him well, however, and taught him to be strong and hardy, and when he was twenty years old she sent him to his parents. He had such an appetite that he eat them out of house and home, and then he made his mother go and beg all over the country till she had enough to buy him three hundredweight of iron to make him a club. Armed with this club, he went forth to seek fortune. In the woods he met a giant carrying a leaden club called *Barbiscat* ('Cat's Beard'), and the two made friends went out together till they met another giant, who carried a wooden club called *Testa di Molton* ('Ram's Head'). They made friends and went out together till they came to a house in a town where magicians lived. The giant with Barbiscat knocked first, and at midnight a magician came out and said, 'Earthworm, wherefore are you come?' then he of Barbiscat was frightened and ran away. The next night the giant with *Testa di Molton* knocked with the same result. But the third night Gian dall' Orso himself knocked, and he

---

[1] I have mentioned the only other wolf-stories that I have met with in the chapter on Excursions round Meran; and at p. 31 of this volume.
[2] Cox's *Aryan Mythology*, vol. i. p. 405.

had no fear, but when the magician came out he knocked him down with many blows of his iron club, and went to fetch the other two giants. When they returned no magician was to be seen, only a trail of blood. They followed the trail till they came to a deep pit, and Zuam dall' Orso made the giants let him down by a rope. In a cave he found the wounded magician and three others besides, by slaying whom he delivered a beautiful maiden. The giants drew her up, but abandoned him. Then he saw a ring lying on the ground, and when he took it up and rubbed it two Moors appeared and asked him what he wanted. 'I want an eagle, to bear me up to earth,' he said. So they brought him a big eagle, 'but,' said they, 'he must be well fed the while.' So he bid them bring him two shins of beef, and fed him well the while, and the eagle bore him to the king; who finding he was the deliverer of his daughter, killed the two giants, and gave him plenty of gold and silver, with which he went back to his home and lived happily and in peace,—a very homely termination, welcome to the mountaineer's mind. In the Lederthal version he was so strong at two years old that he lifted up the mountain under which the bear's den was, and ran back to his mother; but at school he killed all the children, and knocked down the teacher and the priest, and was sent to prison. Here he lifted the door off its hinges, and went to the judge, and made him give him a sword, with which he went forth to seek fortune. With the two companions picked up by the wayside, who for once do not play him the trick of leaving him below in the cave, he delivers three princesses, and all are made happy. In another version, where he is called 'Filomusso the Smith,' and is nurtured by an ass instead of a bear, the provision of meat for feeding the eagle is exhausted before he reaches the earth, and he heroically tears a piece of flesh out of his own leg, and thus the flight can be completed.

2. The following version of the story of Joseph and his Brethren is quaint:—A king had three sons. The two elder were grown up, while Jacob (the Italian is not given) was still quite small, and was his father's pet. One day, when the king came back from hunting, he was quite out of sorts because he had lost the feather (la penna dell' uccello sgrifone) he was wont always to wear. When everyone had sought for it in vain, little Jacob came to him, and bid him eat and be of good cheer for he and his brothers would find the feather. The king promises his kingdom to whichever of the three finds it. Little Jacob finds the feather, and carries it full of joy to his brothers. The brothers, jealous that he should have the kingdom, kill him and take the feather to their father. A year after a shepherd finds little Jacob's bones, and takes one of them to make a fife, but as soon as he begins to play upon it the fife tells the whole story of the foul play. The shepherd takes it to the king, who convicts his two sons, has them put to death, and dies of grief.

3. Here is a homely version of Oidipous and the Sphinx:—A poor man owed a large debt and had nothing to pay it with. The rich man to whom he owed it came to demand the sum, and found only the poor man's little boy sitting by the hearth. 'What are you doing?' asked the rich man. 'I watch them come and go,' replied the boy. 'Do so many people come to you then?' enquired the rich man. 'No man,' replied the boy. Not liking to own himself puzzled, the rich man asked again, 'Where is your father?' 'He's gone to plug a hole with another hole,' replied the boy. Posed again, the rich man proceeded, 'And where's your mother?' 'She's baking bread that's already eaten,' replied the boy.

'You are either very clever or a great idiot,' now retorted the rich man; 'will you please to explain yourself?' 'Yes, if you will reward me by forgiving father his debt.' The rich man accepted the terms, and the boy proceeded.

'I'm boiling beans, and the bubbling water makes them seethe, and I watch them come and go. My father is gone to borrow a sum of money to pay you with, so to plug one hole he is making another. All the bread we have eaten for a fortnight past was borrowed of a neighbour, and now mother is making some to pay it back with, so I may well say what she is making is already eaten.'

The rich man expressed himself satisfied, and the poor man was delivered from the burden of his debt.

4. A poor country lad once went out into the wide

world to seek fortune. As he went along he met a very old woman carrying a pail of water, with which she seemed sadly overladen. The poor lad ran after her, and carried it home for her. But she was an Angana, and to reward him she gave him a dog and a cat, and a little silver ring, which she told him to turn round whenever he was in difficulty. The boy walked on, thinking little about the old woman's ring, and not at all believing in its efficacy. When he got tired with his walking he laid down under a tree, but he was too hungry to sleep. As he lay tossing about he twirled the ring round without knowing what he was doing, and suddenly an old woman appeared before him, just like the one he had helped, and asked what he wanted of her. 'Something to eat and drink' was the ready and natural answer. He had hardly spoken it when he found a table spread with good things before him. He made a good meal, nor did he neglect to feed his dog and cat well; and then they all had a good sleep. In the morning he reasoned, 'Why should I journey further when my ring can give one all one wants?' So he turned the ring round; and when the old woman appeared he asked for a house, and meadows, and farming-stock, and furniture; and then he paused to think of what more he could possibly desire; but he remembered the lessons of moderation his mother had taught him, and he said, 'No, it is not good for a man to have all he wants in this world.' So he asked for nothing more, but set to work to cultivate his land. One day when he was working on his land, a grand damsel came by with a number of servants riding after her. The damsel had lost her way, and had to ask him to lead her back to the right path. As they went, she talked to him about his house and his means, and his way of life; and before she had got to her journey's end they were so well pleased with each other that she agreed to go back with him and marry him; but it was the ring she was in love with rather than with him. They were no sooner married than she got possession of the ring, and by its power she ordered the farm-house to be changed into a palace, and the farm-servants into liveried retainers, and all manner of luxuries, and chests of coin. Nor was she satisfied with this. One day, when her husband was asleep in a summer-house, she ordered it to be carried up to the highest tip of a very high mountain, and the palace far away into her own country. When he woke he found himself all alone on the frightful height, with no one but the dog and cat, who always slept the one at his head and the other at his feet. Though he was an expert climber it was impossible to get down from so sharp a peak, so he sat down and gave himself up to despair. The cat and dog, however, comforted him, and said they would provide the remedy. They clambered down the rugged declivity, and ran on together till they came to a stream which puss could not cross, but the dog put her on his back and swam over with her; and without further adventure they made their way to the palace where their master's wife lived. With some

cleverness they manœuvred their way into the interior,
but into the bed-room there seemed no chance of effecting an entrance. They paced up and down hour by
hour, but the door was never opened. At last, when all
was very still, a mouse came running along the corridor.
The cat pounced on the mouse, who pleaded hard for
mercy in favour of her seven small children. 'If I
restore you to liberty,' said the cat, 'you must do
something for me in return.' The mouse promised
everything; and the cat instructed her to gnaw a hole
in the door, and fetch the ring out of the princess's
mouth, where she made no doubt she kept it at night
for safety. The mouse kept her word, and obeying her
directions punctually, soon returned with the ring; and
off the cat and dog set on their return home, in high
glee at their success. It rankled, however, in the dog's
mind, that it was the cat who had all the glory of recovering the treasure; and by the time they had got
back to the stream he told her that if she would not
give him the satisfaction of carrying the ring the rest
of the way, he would not carry her over it. The cat
would not accept his view, and a fight ensued, in the
midst of which the ring escaped them both and fell
into the water, where it was caught by a fish. The cat
was in despair, but the dog plunged in and seized the
fish, and by regaining the ring earned equal right to
the merit of its recovery, and they clambered together
in amity. Their master was rejoiced to receive his ring
once more, and by its power he got back his homestead
and farm-stock, and sent for his mother to live with

him, and all his life through took great care of his faithful dog and cat; but the perverse princess he ordered the ring to transfer in the summer-house to the peak whither she would have banished him. When all this was set in order he threw away the ring, because he said it was not well for a man to have all his wishes satisfied in this world.[1]

[1] I have thought this one of the best specimen tales, as the two stories of the Three Wishes and the Three Faithful Beasts are leading ones in every popular mythology. I have named a good many variants in connexion with their counterparts in the *Folklore of Rome*, and a more extensive survey of them, together with a most interesting analysis of their probable origin, will be found in Cox's *Mythology of Aryan Nations*, vol. i. pp. 144 and 375. I had thought that these, being strung together in the text version, was owing to a freak of memory of some narrator who, having forgotten the original conclusion of the former story, takes the latter one into it; but, curiously enough, in the note to the last-named page of Mr. Cox's work, he happens actually to establish an intrinsic identity of origin in the two stories. The Three Wishes story has also a strangely localized home in the Oetzthal, which, though properly belonging to the division of North Tirol, I prefer to cite here, for the sake of its analogies. Its particular home is in the so-called Thal Vent, on the frozen borders of the Gletscher described by Weber, as appalling to a degree in its loneliness, and in the roaring of its torrents, and the stern rugged inaccessibility of its peaks. Here, he says, three Selige Fräulein (Weber, like Schneller, translates everything inexorably into German; this may have been an Enguana) have their abode in a sumptuous subterranean palace, which no mortal might reach. They are also called *die drei Feyen*, he says, forming a further identification with the normal legend, but he does not account for the penetration of the French word into this unfrequented locality. They were kind and ancillary to the poor mountain folk, but the dire enemies of the huntsman, for he hunted as game the creatures who were their domestic animals (here we have the nucleus of a heap of various tales and legends of the pet creatures of fairies and hermits becoming the intermediaries of supernatural communication). The Thal Vent legend proceeds that a young shepherd once won the regard of the *drei Feyen;* they fulfilled all his wishes, and gave him constant access to their palace under the sole condi-

The following legend of St. Kümmerniss is very popular in Tirol. Churchill, in his 'Titian's Country,' mentions a chapel on the borders of Cadore and Wälsch-Tirol, where she is represented just as there described, but he does not appear to have inquired into its symbolism. There was once a heathen king who had a daughter named Kümmerniss, who was fair and beautiful beyond compare. A neighbouring king, also a heathen, sought her in marriage, and her father gave his consent to the union; but Kümmerniss was distressed beyond measure, for she had vowed in her own heart to be the bride of heaven. Of course her father could not understand her motives, and to force her to marry put her into a hard prison. From the depths of the dungeon Kümmerniss prayed that she might be so transformed that no man should wish to marry her; and in conformity with her devoted petition, when they came to take her out of the prison they found that all her beauty was gone, and her face overgrown with long hair like a man's beard. When her father saw the change in her he was indignant, and asked what had befallen her. She replied that He whom she adored had changed her so, to save her from marrying the heathen king

---

tion that he should never reveal its locality to any huntsman. After some years the youth one day incautiously let out the secret to his father, and from thenceforth the *drei Feyen* were inexorable in excluding him from their society. He pleaded and pleaded all in vain, and ultimately made himself a huntsman in desperation. But the first time he took aim at one of their chamois, the most beautiful of the three fairies appeared to him in so brilliant a light of glory, that he lost all consciousness of his actual situation and fell headlong down the precipice.

after she had vowed herself to be His bride alone. 'Then shall you die, like Him you adore,' was her father's answer. She meekly replied that she had no greater desire than to die, that she might be united with Him. And thus her pure life was taken a sweet sacrifice; and whoso would like her be altogether devoted to God, and like her obtain their petition from heaven, let them honour her, and cause her effigy to be painted in the church. So many believed they found the efficacy of her intercession, that they set up memorial images of her everywhere, and in one place they set one up all in pure gold. A poor minstrel once came by that way with his violin; and because he had earned nothing, and was near starving, he stood before St. Kümmerniss and played his prayer on his violin. Plaintive and more plaintive still grew his beseeching notes, till at last the saint, who never sent any away empty, shook off one of her golden shoes, and bid him take it for an alms. The minstrel carried the golden shoe to a goldsmith, and asked him to buy it of him for money; but the goldsmith, recognizing whence it came, refused to have anything to do with sacrilegious traffic, and accused him of stealing it. The minstrel loudly protested his innocence, and the goldsmith as loudly vociferated his accusation, till their clamour raised the whole village; and all were full of fury and indignation at the supposed crime of the minstrel. As their anger grew, they were near tearing him in pieces, when a grave hermit came by, and they asked him to judge the case. 'If it be true that the man obtained one shoe by his minstrelsy, let him play till he obtain the other in our sight,' was his sentence; and all the people were so pleased with it, that they dragged the minstrel back to the shrine of St. Kümmerniss. The minstrel, who had been as much astonished as anyone else at his first success, scarcely dared hope for a second, but it was death to shrink from the test; so he rested his instrument on his shoulder, and drew the bow across it with trembling hand. Sweet and plaintive were the shuddering voice-like tones he sent forth before the shrine; but yet the second shoe fell not. The people began to murmur; horror heightened his distress. Cadence after cadence, moan upon moan, wail upon wail, faltered through the air, and entranced every ear and palsied every hand that would have seized him; till at last, overcome with the intensity of his own passionate appeal, the minstrel sank unconscious on the ground. When they went to raise him up, they found that the second golden shoe was no longer on the saint's foot, but that she had cast it towards him. When they saw that, each vied with the other to make amends for the unjust suspicions of the past. The golden shoes were restored to the saint; but the minstrel never wanted for good entertainment for the rest of his life.

'Puss in Boots' figures in the Folklore of Wälsch-Tirol as 'Il Conte Martin della Gatta;' its chief point of variation is that no boots enter into it at all, otherwise the action of the cat is as usual in other versions.

There is another class of stories in which the townspeople indulge at the expense of the uninstructed peasants in outlying districts, and which their extreme simplicity and naïveté occasionally justify. I must not close my notice of the Volklore of Wälsch-Tirol without giving some specimens of these. It may be generally observed that stories which have no particular moral point, and are designed only to amuse without instructing, are as frequent in the Trentino as they are rare in the German divisions of Tirol.

Turlulù[1] was such a simple boy that he could not be made to do anything aright; and what was worst was, he thought himself so clever that he would always go off without listening to half his instructions. One day his mother sent him with her last piece of money to buy a bit of meat for a poor neighbour; 'And mind,' she said, 'that the butcher doesn't give you all bone.' 'Leave that to me!' cried Turlulù without waiting for an explanation; and off he went to the town. The butcher offered him a nice piece of leg of beef. 'No, no, there's bone to that,' cried Turlulù; 'that won't do.' The butcher, provoked, offered him a lump of lights. Turlulù seeing it look so soft, and no bone at all to it,

---

[1] They are called 'Lustige Geschichte,' 'Storielle da rider.' The Germans have a saying that 'in jede Sage haftet eine Sache;' the 'Sache' is perhaps more hidden in these than in others. I have pointed out counterparts of the following at Rome and elsewhere in *Folklore of Rome*.

went off with it quite pleased, but of course the poor neighbour had to starve. When his mother found what he had done, she was in great distress, for she had no money left; so she sent him with a piece of homespun linen to try to sell it. 'But mind you don't waste your time talking to gossiping old women,' she said. 'Leave that to me, mother,' cried Turlulù; and off he ran. As he got near the market-place, he began crying, 'Fine linen! who wants to buy fine linen!' Several countrywomen, who had come up to town to make purchases, came to look at the quality. 'Go along, you gossiping old things; don't imagine I'm going to sell it to you!' cried Turlulù, and he ran away from them. As he ran on he saw a *capitello*[1] by the wayside. When he saw the image of the Blessed Virgin, looking so grave and calm, he said, 'Ah, you are no gossip, you shall have my linen;' and he threw it at her feet. 'Come, pay me!' he cried presently; but of course the figure moved not. 'Ah, I see, you've not got the money to-day; I will come back for it to-morrow.' When he came back on the morrow the linen had been picked up by a passer-by, but no money was forthcoming. 'Pay me now,' said Turlulù; but still the figure was immovable. Again and again he repeated the demand, till, finding it still unheeded, he took off his belt, and hit hard and fast upon the image. So great was his violence, that in a very short time he

---

[1] *Capitello*, in Wälsch-Tirol, is the same as *Bildstöcklein* in the German provinces—a sacred image in a little shrine.

had knocked it to the ground; and lo and behold, inside the now uncovered pedestal were a heap of gold pieces, which some miser had concealed there for greater security. 'My mother herself will own this is good pay for the linen,' cried Turlulù, as he filled his pockets, 'and for once she won't find fault.' His way home lay along the edge of the pond, and as he passed the ducks were crying, 'Quack! quack! quack!' Turlulù thought they were saying *Quattro*, meaning that he had four pieces of gold. 'That's all you know about it,' cried Turlulù; 'I've got many more than four, many more.' But the ducks continued to cry 'Quack.' 'I tell you there are more than four,' reiterated Turlulù impetuously, but the ducks did not alter their strain. 'Then take them, and count them yourselves, and you'll see what a lot there are!' So saying, he threw the whole treasure into the mud; and as the ducks, scared by the noise, left off their 'quack,' he satisfied himself that he had convinced them, and went home to boast to his mother of the feat.

A showman came through a village with a dancing-bear. The people went out to see him, and gave him plenty of halfpence. 'Suppose we try our luck, and go about showing a bear too; it seems a profitable sort of trade,' said one of the lookers-on to another. 'Ay, but where shall we find one?' objected the man addressed. 'Oh, there must be bears to be found; it needs only to go out and look for them.' They went out to look for a bear, and at last really found one,[1] which ran before them and plunged into a cave. 'I'll tell you what we'll do,' said the peasant who had proposed the adventure, 'I'll creep into the cave and seize the bear, and you take hold of my legs and pull us both out together.' The other assented; and in went the first. But the bear, instead of letting him seize it, bit off his head. The other pulled him out as agreed, but was much astonished to find him headless. 'Well, to be sure!' he cried, 'I never noticed the poor fellow came out this morning without his head. I must go home and ask his wife for it.' So saying, he ran back to the man's house. 'I say, neighbour,' he cried, 'did you happen to notice, when your husband went out this morning, whether he had his head on?' 'I never thought to look, replied the wife, 'but I'll run up and see if he left it in bed; but tell me,' she added, 'will he catch cold for going out without his head on?' I don't know as to that,' replied the man; 'but if he should want to whistle he might find it awkward!'

A woman working in the fields one day saw a snail, which spread out its horns as she looked at it. In great alarm, she ran to the chief man of the parish, and told him what she had seen. He, too, was horribly frightened, but he mastered his fear, as became the dignity of his office. In order to provide duly for the safety of his village, he sent two trustworthy men

[1] Bears exist to the present day in Tirol. Seven were killed last year. A prize of from five-and-twenty to fifty florins is given for killing one by various communes.

with a large sum of money to Trent, to buy a sharp sword; and till their return placed all the able-bodied men on guard. When the man brought the sharp sword back from Trent, he called the heads of the Commune together, and said to them: 'I will not exercise my right of sending any of you in peril of his life, but I ask you which of you is ready to encounter this great danger, and whoever has the courage shall receive a great reward.' Hereupon two of the most valiant came forward as volunteers, and were invested with the sharp sword. In solemn silence they marched boldly to the field where the snail was, and they saw him sitting on the edge of a rotten leaf; but at the moment when they had screwed up their courage to smite him with the edge of the sword, the breeze blew down the leaf and the snail with it. They, however, thought the snail was preparing to attack them, and ran away so fast that they tumbled over the edge of an abyss.

The people of a certain village were envious because the church tower of the neighbouring village was higher than theirs. So they held a council to consider what remedy they could apply. No one could think of anything to propose, till the oldest and wisest of them at last rose and advised that a great heap of hay should be laid by the side of their tower, so that it might eat and grow strong, and increase in height. The counsel was received with applause, and every one cheerfully brought his quota to the common sacrifice, till there was a mighty heap of hay laid at the base of the church tower. All the horses and asses that went by, finding such a fine provision of provender laid out for them, ate the hay; but the people seeing the heap diminish, were quite satisfied, and said, 'Our tower must be beginning to grow, you see how fast it eats!'

In Wälsch-Tirol the graves are not decked with flowers on All Souls' Day, as in Germany, but on the other hand it is customary for the parish clergy to gather their flocks round them, and say the Rosary kneeling amid the graves. Doles of bread, locally called *cuzza*, and alms, are given away to the poor on that day, and in some places a particular soup made of beans. The symbolism was formerly carried so far, that these alms, devoted to the refreshment of the souls of the departed, were actually laid on the graves, as if it was supposed that the holy souls would come out and partake of the material food. And thus some even placed vessels of cold water as a special means of solace from their purgatorial pains.[1] In the north of Italy, the feast of Sta. Lucia (December 13) holds the place of that of St. Nicholas among children in Germany; in Wälsch-Tirol the children have the advantage of keeping *both*.

[1] A distinct remnant of Etruscan custom. It is singular, too, that Mr. I. Taylor finds 'faba' to have been taken by the Romans from the Etruscans for a bean, but though the custom of connecting beans with the celebration of the departed is common all over Italy, I do not think the Etruscans provided their dead with beans except along with all other kinds of food (*supra* p. 130-1 *note*).

In Val Arsa, part of the loaves baked on Christmas Eve are kept, as Cross-buns used to be among us. In Folgareit they have a curious game for Christmas-tide. A number of heaps of flour, according to the number of the household, are arranged on the table by the father of the family, some little present being covered up in each; when they are thus prepared the family is admitted, and the choice of places decided by various modes of contest. In several parts, particularly in the Rabbithal, the Lombard [1] custom prevails of putting a huge log on the fire, called the *Zocco di Natale* and the *Zocco di ogni bene*, that it may burn all night and keep the Divine Infant from the cold. The idea, more or less prevalent all over Christendom, that beasts have the gift of speech on Christmas Eve, prevails here no less. A story is told of a peasant who determined to sit up and listen to what his oxen said. 'Where shall we have to go to-morrow?' he heard one say. 'We shall have to fetch the boards for our master's coffin,' replied his companion. The man was so shocked, that he went to bed and died next day. Animals are blessed on St. Anthony's day (January 18), as in Rome.

Carnival is celebrated with representations par-

---

[1] The little book of *Costumi* spoken of above, mentions the 'Zocco del Natale' as in use also in Lunigiana; it is generally of olive-tree, and household auguries are drawn from the crackling of leaves and unripe berries. It cites a letter of a certain Giovanni da Molta, dated 1388, showing that the custom has not undergone much change in five hundred years.

taking somewhat of the character of 'Passion Plays,' though always with more or less humorous treatment of their subject. Till lately there lingered a curious pastime at this season, in which on Giovedì grasso there was a contest, according to fixed rules, between the masked and unmasked inhabitants, for certain cakes (*gnocchi*) made of Indian corn, whence the day is still called *Giovedì dei gnocchi*. It commemorated a fight between the men of Trent and them of Feltre, who tried to carry off their provision while they were building the walls of Trent, in the time of Theodoric King of the Visigoths. S. Urban is considered the patron of vineyards in Etschland, and on his feast his images are hung with bunches of grapes.

Here are a few specimens of their popular sayings and customs. When it thunders the children say, *Domeniddio va in carozza*. The chirping of a cricket, instead of being reckoned a lucky token, forebodes death. Sponsors are regarded a person's nearest relations, and at their funeral they go as chief mourners before all others. Marriages in May are avoided. The reason why the bramble always creeps along, instead of growing erect, is, because once a thorny bramble branch caught the hair of the Blessed Virgin; before that it grew erect like other trees. Cockchafers are blind, because one of them once flew into the Blessed Virgin's face and startled her; before that they had sight. Swallows are called *uccelli della Madonna*, but I have not ascertained the reason. Scorpions, which are venomous in Italy, are not so in the Italian Tirol, because one fell once into St. Vigilius' chalice at Mass. I will conclude with some popular riddles, showing a traditionary observation of the movements of the heavenly bodies, but not much humour:

> Due viandanti,
> Due ben stanti,
> E un cardinal?[1]

> Gh' è 'n prà
> Tutto garofalà:
> Quanca se vien el Papa con tutta la sô paperia
> En garòfol sol no l'è bon de portar via?[2]

> Piatto sopra piatto,
> L'omo ben armato,
> Donna ben vestita.
> Cavalleria ben fornita?[3]

C'è un palazzo, vi son dodici camere, ognuna ne ha trenta travi, e vi son due che si corrono sempre l'uno dietro all' altro e non si raggiungono mai?[4]

---

[1] Two travellers, two prosperous ones, and a cardinal?—*Answer.* Sun and moon; earth and heaven; and the ocean.

[2] There is a meadow overblown with carnations, yet if the Pope came with all his court, not one solo carnation would he be able to carry off?—*Answer.* The heaven beaming with stars.

[3] Plate upon plate; a man fully armed; a lady well dressed; a stud well appointed?—*Answer.* Heaven and earth; the sun; the moon; the stars.

[4] There is a palace with twelve rooms; each room has thirty beams, and two are ever running after each other through them without ever catching each other?—*Answer.* The palace is the year, the rooms the months, the beams the days, and day and night are always following each other without overlapping.

O mein Tirol! wie ich mit Schmerzentzücken
Dich nun geschaut vor meinen feuchten Blicken.
So lebt dein rührend Bild im tiefsten Sinn.
Nimm denn, Tirol, des Schmerzbegeistrungstrunk'nen,
Des ganz in dich Verlornen und Versunk'nen
Liebvolles Lebewohl, mit Liebe hin!

EDUARD SILESIUS.

INDEX.

AAR

AARAU, 39
 Abel, Arnold, and Bernhard, 251
Abraham's Garden, 401
Absam, 157–61
Acatius, St., legend of, 37
Ache, Brandenberger, 126
— Gebiet der grossen, 64
Achen Pass, 133
Achenrain, 127
Achensee, 133–4
Achenthal, 36, 132 et seq.
— village, 132, 133
Adige. See Etsch
Ainliffen, 55
Ala, 341
Albuin, St., 356–7
Altbach, 112
Altenstadt, 14, 44
Altissimo di Nago, 406
Altrans, 219
Altrei, 375
Ambras, 234, 259–69
Ammergau, 34
Ampass, 219
Andalusiten, 331
Andreas of Rinn, St., 210, 211 et seq.
Angana, 413, 425–8

ANG

Angels assist in building churches, 34, 405
Annaunia, 358, 361
Anterivo, 375
Anthony, St., 256, 438
Appenzel, mountains of, 17, 19, 42
Archenthal, 26
Ardetzen, the, 13, 42
Arlberg, the, 29, 30
Arthur, King, on Maximilian's monument, 251–2
Artists of Tirol, 36. 84, 86, 112, 136–7, 144–5, 146, 151, 160, 180, 205, 206, 211, 216, 248 et seq., 274, 276, 281 and note, 290–1, 325, 339, 354–5, 376, 378, 384–5, 395
Arz, 366
Asam, Kosmas Damian, 290–1
Aschbach, 207
Asgard, 4
Ass-boy, 423
Au, xvii note, 24, 111
Auferstehungsfeier, 298
Auffach, 112
Auflängerbründl, 66
Avio, 342
Avisio, 341, 357, 381
Axams, 334, 337

BIA

BAIERISCHE-RUMPEL, 286–8
Baldur, 4
Baptism, children raised to life for, 27–8, 355
Barbara, St., legend of, 37
Baselga, 382
Bauern-Comödie, 214, 239–40, 257, 346, 438–9
Baumkirchen, 147
Bavarians in Tirol. 16, 74, 111, 150–1, 177–8, 286–8, 300 et seq., 324, 325–7, 332
Beans, 424, 437 and note
Bears in Tirol, 16, 31, 165–6, 420 et seq., 434–5
Beatrik, 413 and note
Bedelmon, 413
Befreiungskämpfe, 310 –1, &c.
Bolasis, Schloss, 359
Bells rung without hands, 165–6, 218
Berchtl, 117. 124, 417
Berg Isel, 156, 300
Bernardo, St., 363
Bertha, 417
Beseno, 342
Bezzecca, 400
Biagio, St., 367

444 INDEX.

BIB

Biberwier, 32
Bienerweible, the, 122-4, 312
Birds in Tirolean mythology, 87, 135, 141-2, 196-8, 325, 422-3, 434, 439
Birgitz, 337, 338
Blase, St., legend of, 38
Blasienberg, 334
Blendsee, 33
Blind, Karl, 3 et seq., 214 and note
'Blood, relics of the holy,' 139, 324-5
Bludenz, 25, 47, 48, 238
Bodeusee, 17, 20, 39-40, 45
Borghetto, 341
Borgo di Val Sugana, 392
Bormio, 31
Bosentino, 388
Brandenbergerthal, 127[1]
Brava Berta, 417-8
Bregenz, 17-9
Breitenwang, 35
Brenner Pass, the, 246, 247, 278, 287
Brenta, the, 382, 386, 392
Brentonico, 407
British missionaries in Tirol, 23, 43
Brixenthal, 64
Brixlegg, 79, 112, 121, 124
Brothers' strife, legend of, 391-2
Bruederhuesle, the, 27-8
Bruneck, 247
Buch, 201
Buchau, 133
Buchs, 13, 23
Burgleckner, 121-2
Buried cities (and submerged), legends of, 26, 133, 204, 342-3

BUR

Burning hand, legend of, 167
Busa del Barbaz, 343
Busetti, Christopher, 368

CADERZONE, 405
— Cadine, 356, 404
Cadore, 429
Caldonazzo, lake, 386
— village, 391
Calliano, 249
Campiglio, 405
Campi neri, 361-3
Campolongo, 387
Canal San Bovo, 398-9
Caravaggio, Madonna di, 382
Caresolo, 405
Carinthia, 247, 286, 297
Carneval, 99, 100, 346, 393-4, 419, 438-9
Castel Alto, 396
— Barco, 342
— del Dosso Maggiore, 407
— Fondo, 366
— Madruzz, 400
— Nuovo, 344
— Pietra, 249
— Rotto, 396
— Telvana, 393-4
— Tesino, 397
Cats in Tirolean mythology, 110, 425, 431-2
Cauria, 399
Cavalese, 375-7
Cavedine, 401
Cembra torrent, 357
— village, 375
Centa torrent, 391
Chamois in Tirolean mythology, 197, 318, 428-9

CYR

Charlemagne, 254
— legend of, 18-9
Charles, St., Borromeo, 22, 209 et seq., 363-4
— Prince of Tirol, 285
— Quint, 190, 245, 246-7, 250, 259, 263, 364, 405
— VI., Emperor of Germany, 289
Chinese legend, 131
Christberg, the, 27
Christian I. of Denmark, 243
Christina of Sweden, 256 et seq., 280, 283
Christmas customs, 100, 438
Christof, St., 29-30
Cima d'Asta, 396
Claudia Felicità, 285
— de' Medici, 122, 219, 277, 282, 283, 326
Cles, 360-4
Cloz, 366
Cobweb paintings, 308
Cockchafers, 439
Cogolo, 365
Cold torment, the, 129
Colin, Alex. 180, 251 et seq., 269, 299
Columban, St., 43, 345
Condino, 405
Constance. See Bodensee
Corno de'tre Signori, 365
Costalta, 396
Costumes, 44, 45, 249, 278, 385
Cow-fighting, 98-9
Cox, Rev. G. W., 2, 3, 119, 120, 157, &c.
Crickets, 439
Cristofero, San, 387
Cyriacus, St., legend of, 38

---

[1] The simplicity of the people of this valley is celebrated in many 'Men of Gotham' stories.

INDEX. 445

DAL

DALAAS, 27
  Dambel, 371
Damenstift, 20, 209, 294, 295
Damüls, 25
Dance of Death, 36
Dancing in Tirol, 98
Dandolo Tullio, 8
Daniel, Prophet, patron of miners, 201
Dante on popular traditions, 8
— in Tirol, 342
— quoted, 129, 265
Dasent, Dr., 129, 267, &c.
Deaf made to hear, 355
Death as a maiden, 390-1
Denno, 359
Devil in Tirolean mythology. 25, 35, 77, 82, 97, 233, 266, 351-4
Devil's house, 48-9, 350
Devonshire legend of Bertha, 418
Dialects of Tirol, xvi-xix, 28, 66, 79, 111, 202, 330, 334, 354, 371, 406, &c.
Dietrich von Bern, 252, 413 note, &c.
Divisions of Tirols, xiv-v
Dirstenöhl, 327
Dispensations to bishops to marry, 277, 284, 401
Doggi, 410-1
Dogs in Tirolean mythology, 425
Dolomites, 379, 408
Domenichino, 396
Donna Berta, 417-8
Dormitz, 32

DOR

Dornbirn, 20
Dos Trento, 356
Dragon, 35, 157 note, 233
Drei Herren Spitz, 365
Dreyling, Hanns, 179-80
Dumb made to speak, 223, 388
Dürer, A., picture by, 146, 151
Duxerthal, 88-90, 202
Dwarfs, 113-4

EASTER, 4
  Eben, 71, 73, 135-6
Eguane, 413
Ehreguota, 19
Ehrenberger Klause, 34, 286
Eleanor of Scotland, 146, 243
Elias, 404
Elmo, Sant', 38
Embs, 21
Engadeiners, the, 58
English officers in Austrian service, 327
Enguane, 413, 428
Enneberg, the, 297
Epiphany customs and legends, 99-101, 116 et seq.
Erasmus, St., legend of, 38
— Quillinus, 151
Ernst der Eiserne, 84, 240, 252
Erzherzog, Johann, 300
Etruscan remains in Tirol, xiv note, xvi-xix, 79, 129-31 note, 307, 346-7, 350, 371-4, 407 et seq., 416 note, 417 note, 438 note [1]
Etsch, 74, 249, 341

FEY

Etschthal, 340 et seq., 358, 439
Euganeans, 341, 413, 425, 428
Eusebius, St., 14 et seq.
Eustachius, St., legend of, 38
Evasthal, 374
Executions in Tirol, 163

FAITHFUL beasts, 425-8
Falkenstein, 200, 202
Fare, travellers', the lesson of, 171-2
Fassathal, 357, 374, 379, 420
Fatscherthal, 330
Federbett, the, 90
Feldkirch, 12 et seq., 22, 24, 43 et seq., 298
Felsenau, 24
Ferdinand I., 80, 163, 245 et seq., 324-5
— II., of Tirol, 209, 258 et seq.
— II., Emperor of Germany, 277
— I., Emperor of Austria, 281, 305
— Karl, 123, 160, 280, 282-3, 312
Ferdinandeum, the, 306-8
Ferklehen, 329
Fern, auf dem, 33
— Pass, 247
— See, 33
Ferneck, 335
Fersinathal, 385-6
Feuriger Mann, 167
— Verräther, the, 379
Feyen die drei, 428-9 note

---

[1] Dennis, *Cities and Cemeteries of Etruria*, vol. 1, pp. xxxiv-v, mentions the Etruscan remains that had been found at Mattrey (of which he gives a cut) and other places in Tirol up to his time.

FIA

Fiabe, 412, 420 *et seq.*
Fiddler, the prevailing, 429-31
Fidelis, St., 13
Fiecht. *See* Viecht
Fiegers, the, 84, 151, 173, &c.
Fife. *See* Pipe
Fig-coffee, 171
Filò (filatorium) 33, 396, 416-9 *et seq.*
Filomusso the Smith, 423
Finstermünz, 31, 287
Fire-worship, 346
Flatz, 17
Flavone, 359
Fleims, Fleimserthal, 74, 357, 398
Floitenthal, 88
Flowers in Tirolean mythology, 37, 70-1, 195, 216, 405
Folgareit, 438
Fondo, 366
Pragenstein, 324
Francis I., Emperor of Germany, 164, 292
— II., Emperor of Germany, 296
Franzensbrücke, 29
Franzosenbühl, 356
Frastanz, 24
Frauenberg, 14-16
Frau Hütt, 311-2
Freia, 4, 5, 6, 110
Freihof, the, 241
French in Tirol, the, 160, 177, 286 *et seq.*, 296, 326, 356, 357, 359, 381, 395-6, 397
Frescoes, curious, 17, 31, 186, 243, 364
Freundsberg, 188 *et seq.*
Freyenthurn, 365
Fridolinskapelle, 14-5
Friedberg, 207

FRI

Friedrich mit der leeren Tasche, 26, 75-8, 84, 96, 207, 235, 237, 238-43, 253, 337, 393
Fritzens, 148
Froberte, 417-9
Frogs, 339
Fulpmes, 339
Fügen, 84, 85
Fuggers, the, 259 *et seq.*
Fuggers, the, in Tirol, 80, 173, 350-4
Fussach castle, 20
Füssen, 32, 34, 254

GALL, St., 43
Gallwiese, the, 334
Galzein, 201-2
Gampen U. l. Frau, a. d., 366
Gannes, 408 *et seq.*
Garibaldi repulsed from Tirol, 400
Garnets in Tirol, 87-8, 397
Gebhartsberg, St., 17
Gefrorene Wand, the, 88, 91
Georgenberg, St., 80, 137 *et seq.*
Gerlos. 79, 87
Germanic mythology, 4 *et seq.*, 416, &c.
Gerold, St., 16
Geroldsbach, 334
Giacomo, San, 407
Gian dall' Orso, 420 *et seq.*
Giants, 231, 335, 421
— called *Salvan* in Wälsch-Tirol, 420
Gilgen, St. (Giles), legends of, 37
Giovanni da Udine, 345
Giudicaria, 341, 400 *et seq.*
Giuliano, St., 404

HEI

Glockenhof, the, legend of, 161-3
Glunggeser, 206, 212
Gnadenwald, 147, 161
Goaslahn, 195
Godl, 253
GoldeneDachl-Gebäude, 240-3
Gömacht, 101
Gotham, men of, stories, 432-7
Gotzens, 337
Götzis, 20
Götzneralp, 334
Grafmarterspitz, 206
Gregorian calendar, introduction into Tirol, 273
Greifenstein, 241
Gries, 330
Grigno, 397
Grins, 30
Grintzens, 337
Grödnerthal, 108, 338
Grolina, 393
Grünegg, 161
Guarda, 382
Guarinoni, Hippolitus, 209
Gumpass, 164
Gunglthal, 32
Gutenberg, Schloss, 23

HAINZENBERG, 86-7
Hall, 47, 149 *et seq.*
Hanzenheim, 206
Häring, 132
Harlesanger, 134
Haspinger, 308-9
Haymon, 231 *et seq.*, 335
Heilige Blutskapelle, 324-5
Heiligenkreuz, 164, 202
Heiligenwasser, 222-4
Heinrich das Findelkind, 30
Heiterwang, 34

INDEX.

HEL

Hel, 4, 129 *note*
Henry II., Emperor of Germany, 64–5
— VIII. of England on Maximilian's monument, 250
— of Rottenburg, 67 *et seq.*
— VI. of Rottenburg, 75 *et seq.*
Hilariusbergl, 126
Hildegard, 18–9
Hinter Judicarien, 400
History of Tirol, 231, 236 *et seq.*, 346
Hoch Gerach, 16
Höchst St. Johann, 20
Hofer, Andreas, 156, 300–5, 308
Hohenembs, 20-2
Hörbrunn, 112
Horseshoes, 136 and *note*
Hosennaglòr, 98
Host as talisman, 221-2
Hottingen, 313
Hugh of Lincoln ballad, 217–8
Hulda, 87
Hundskapelle, 315
Hundskehlthal, 88

ILGA, St., legend of, 20
Illthal, 13, 24
Illumination, 346
— of MSS., 306
Imst, 32
Ingenuin's garden, St., 356–7
Innrain, 334
Inn river, 31, 84, 111, 126, 230, 236, 237, &c.
Innsbruck, 225 *et seq.*
— its character, 226 *et seq.*
— hotels, 228–9
— mountains surrounding it, 229

INN

Innsbruck streets, 230
— suburbs, 230 *et seq.*, 259, 269, 278, 310 *et seq.*
— history, 226, 230 *et seq.*
— Goldene Dachl-Gebäude, 240–3
— the Burg, 244, 246, 282, 293, 301
— churches, 247 *et seq.*, 257, 272, 274, 275, 280, 290–1
— Franciscan church, 247
— Hofkirche, 247, 257
— Maximilian's monument, 248 *et seq.*
— Silver chapel, 254, 256, 269–70
— Jesuitenkirche, 257
— the plague, 258, 276, 282
— the Siechen-haus, 258, 276
— earthquake, 270, 281
— Capuchin church, 272, 274
— Servite church, 274
— Jesuit church (Dreiheiligkeitskirche), 275
— Hopfgarten, 278
— Kranach's Mariähülfsbild, 279–81 and *note*.
— Mariähülfskirche, 280
— Pfarrkirche, 236, 280
— University, 285, 294, 296, 300, 305
— the Annensäule, 289
— Landhaus and Gymnasium, 289
— Church of St. John, Nepomucen, 291
— The Triumphpforte, 294

JUD

Innsbruck, the Gottesacker, 299
— The St. Veitskapelle, 299
— Berg Isel, 300–1
— Hofer's monument, 304–5
— Museum (Ferdinandeum), 306–8
— Schiess-stand, 308-9
— population, 309
Innerlahn, 206
Isarthal, 324, 325
Ischgl, 30
Isel Berg, 156, 300, 336
Iselthal, 102
Italians in Tirol, 122–4, 199, 270, 283, 284, 299, 340, 345, 350, 365
Ivano, 397

JACK O' LANTERN, 413
Jenbach, 79, 80, 92–3, 125, 132–3
Jew, Wandering, 117
Jews in Tirol, 22, 212–8, 348, 354–5, 405
Joder, St., legend of, 25
Jodok, St., 254, 334
Johann Erzherzog, 300
Johannissegen, St., 104, 109
John XXIII., 26
Joseph, St., peculiar devotion to, 27–8
Joseph I., Emperor of Germany, 289
— II., Emperor of Germany, 25, 210, 281, 293, 294, 347, 355
— and his brethren, 423
Jötun-fires, 214 *note*
Joubert, in Tirol, 160, 296
Judas, 214

JUD

Judenstein, 210, 212
Judgments. sudden, legends of, 82, 88-91, 120, 123, 133-5 and note, 204, 220-1, 312, 379, 380-1,
Judicarien, 341, 400 et seq.
Julian, St., 404-5
Julius II., 256
Jussulte, 35
Juvenau, 331

KAISERBRUNNEN, 36
Kaisersäule, 164
Kaltern, 284
Karl Philipp, 289
Käsbach, the, 125
Käsbachthal. 132
Kelchsau, 112
Kellerspitze, the, 202
Kematen, 333
Kirchbübel, 132
Klosterthal. 24
Knappen, 113-4, 174 et seq., 201 et seq., 320, 322, 385-6
Kobach, 112
Kolsass, 202, 205, 207
Kramsach, 126
Kranach's Lucas, Madonna, 186, 192, 257, 279-81, legend, 281 note.
Kranebitten, 314
Kreidenfeuer, 287
Kreuz-kapelle, 325
Kropfsberg, 79
Kufstein, 53 et seq., 286, 288
Kugelmoos, 260
Kümmerniss, St., 429-31
Kundl, 64, 78, 92, 110-2

LADYBIRD, legend of, 5

LAG

Lagerthal. *See* Val Lagarina
Lago di Garda, 341, 400
Lafraun. 391
Lähn, 34
Lakes, 25, 33, 34, 119-20, 126, 127, 132, 206, 220, 325, 386, 391, 397, 401, 402
Lampsenjock, the, 140
Landeck. 31-2, 48, 239-40, 287-8
Landmark, legend of moving, 116
Lans, 219-20
Laste, Madonna alle, 354-5
Lavarone, 391
Lavis, 357, 375
Lederer, Paul. 274
Lederthal, 400, 420, 422
Leermoos, 33-4
Legends, the use of, 1 et seq.
Legends, perpetuation of. 9 et seq., 22, 118-21, 158-9, 416-7, and note, &c.
— modern, 13, 32-3. 166. and *note*, &c.
— their extinction, v-ix
Legion, the thundering, 235
Lehner, 315
Leiten, 324
Lendenstreich, Hanns, 254
Lengenthal, 241
Lenothal and Schlucht, 344, 345
Leonhard, St., auf der Wiese, 64-6, 111
Leopold I., Emperor of Germany, 284, 289
Leopold II., Emperor of Germany, 292-3, 295, 296
Leopold V., Arch-Duke, 122, 277, 325

MAG

Lettered lilies, 216 *note*
Letz, 32
Leuchtenburg, 70, 74
Levico, 392
Lichtenstein, 23, 25 *note*
Lichtenthurm, 314
Lichtwer, 79
Liebenberger, 337
Lindau, 17, 45
Lisenthal and Ferner, 331
Literature, learning, & peaceful arts, cultivation in Tirol, 76-8, 257, 259, 273, 306-7
Livo, 361, 363
Lizumthal, 206, 338
Lizzana, 342
Löffler, 180, 253, 313
Loki, 3, 411
Lombard customs. 438
Lorenzo, St., of Brindisi, 275
Lorenzo, San, 396
Loreto-Kirche, 163-4, 402
Lorg, 408, 411
Lothair the Saxon, 35
Lucius, St., 23
Luigi del Duca. 249
Lunigiana, 417 *note*, 438 *note*
Lustenau, 20
Luterns, 25
Lutherans in Tirol. 85 175-6, 245, 247, 248, 277
Luziensteig, 23

MADONNA DI CARAVAGGIO, 383
— alle Laste, 354-5
— del Monte, 344
— di Pinè, 382
Madruzz, 400-2
Magdalenenbründl, 331
Magras, 363

INDEX.   449

MAL
Malachite, 200
Male, 264
Mangtritt, 34–5
Margaret St., in Lagerthal, 344
Margareth, 201
Margarethenkapf, 13
Margaretha Maultasch, 30–1, 54, 237, 263
Maria Bianca, 243, 244–5, 250, 252
Mariähülfsbild. *See* Kranach
Maria-Larch, 147
Mariä-Rastkapelle, 87
Maria-Stein, 131
Mariathal, 126–7
Maria Theresa, 257, 275, 281, 291
Marmolata, the, xi, 379–81
Marriage customs, 102–10, 378–9, 386, 439
Martin, St., 147
Martinswand, the, 308, 315, 316–23
Mary of Burgundy, 243, 249, 253
Massena in Tirol, 298
Massenza, Sta., 401
Massimeno, 405
Matthias Corvinus, 318, and *note*
Mattrey, 89, 130 *note*, 339, 445 *note*
Maturi, Antonio, 364
Matzen, 79
Maurice, Elector of Saxony, 247
Maximilian, Emperor, 54–5, 74, 126, 146, 152, 173, 215, 226, 243, 244–5, 248, 316 *et seq.*, 355
— his acts, 249–50 *note*; his monument, 248, 281 *et seq.*
— the Deutschmeister, 273–7

MAX
Maximilian, Elector of Bavaria, 286 *et seq.*
Max-Höhle, 322
Mayrhofen, 79, 87
Medicinal and miraculous springs, 22, 23, 112, 164, 207, 219, 222–4, 310, 313, 333, 363, 396, &c.
Meinhard II., 230
Melach, 330, 331
Melans, 157
Merboth, Diedo and Ilga, 20
Mezzana, 364
Mezzo Lombardo, 358
Michael, St., 147
Michel, San, 357
Mils, 161, 207–8
Milser, Oswald and Ottilia, 194 *et seq.*
Mines and mining legends, 27, 78, 82–3, 112, 113–5, 146, 173 *et seq.*, 200 *et seq.*, 324, 385–6
Mines gold, of the Bishop of Brixen, 86
Minstrel, the prevailing, 429–31
Mitschnau. *See* Wildschönau
Mittersee, 33
Mitterwald, 34
Moena, 379
Moll, 256
Möls, 206
Mommsen on 'tavola clesiana,' 361
Montafonthal, 24, 26
Montanaga, 382, 383
Mooserthal, 126
Mori, 343, 407
Mortaso, 403
Mühlau, 254, 254–6, 310 *et seq.*
Müller, Professor Max, 2, 119, 158, &c.

OER
Münster, 126
Mutters, 336
Myth-collectors, Tirolean, xii–xiv
Mythology, Germanic, interest of Englishmen in, 3 *et seq.*
Myths. *See* Legends

NAGO, 406
Nassereit, 32–3
Natters, 336–7
Nauders, 341
Navisthal, 206
Neumarkt, 375
Neurätz, 331
Neustift, 339
Ney, Marshal, in Tirol, 300
Nibelungenlied, Tirol's place in, 358
Nick, 408 *et seq.*
Niedermich, 111
Niederau, 112
Nine, the year, 147, 301, 406
Noce torrent, 366
Nockspitze, 337
Nonsthal and Nonsberg, 165, 358
Nork, Norg, &c., 408 *et seq.*
Nossa, Signora del feles, 388
Nothburga, St., legends of, 66 *et seq.*, 71, 73, 135, 210
Novella torrent, 366

OBERAU, 111
Ober Perfuss, 333
Oberriet, 12, 40, 52
Oberinnthal, 29, 287
Odin. *See* Wodan
Oerkelen, 410

G G

INDEX.

OET

Oetzthal, 101, 239, 331, 338, 428
Orco-myths, 388-9, 408 et seq.
Orgen, 410
Ogre, 410
Orto d'Abraham, 402
Ospedaletto, 397
Ostara, 4
Oswald, St., 325
— Milser, 194, 324-5
— von Wolkenstein, 337-8
Ottoburg, 236-7
Ottokapelle, the, 63
Oxen in Tirolean mythology, 73, 438
Ozolo, 365

P ANCHIÀ, 398
Pantaleone, St., legend of, 38
Puracelsus, Theophrastus, 265 et seq.
Passes, 24, 29, 31-2, 133, 247, 275, 395, 399-400, &c.
Passeyerthal, 303, 410
Passion-plays, 214, 346, 438-9. See also 'Bauern-Comödie.'
Patscherkofl, 337
Patznaunthal, 30
Pejo, Val di, 364
Pellizano, 364
Pelugo, 405
Pergine, 385
Pergola, 389
Pertisau, 132
Perugino, 348
Pestschutzheiligen, 276, 343, 406, &c.
Petroleum in Tirol, 327
Pfäffers, 13, 23, 277
Pfannenberg, 17-8
Pflaun, 359
Pfunerjoch, 206
Philippine Welser, 261 et seq., 308

PHO

Phol, 4
Pigs in Tirolean mythology, 72, 312 note
Pilate's wife, 113
Pill, xviii, 202
Pinè, pilgrimage of, 382-5
Pinzolo, 406
Pioneers previous in Tirol, x-xi, 398
Pipe, in Tirolean mythology, xviii, 423
Pirschenheim, 333
Pitzthal, 32
Pius VI., 294, 295, 338
Plausee, 34
Plaknerjoch, 112
Pontlatzerbrücke, 287
Porta Klaudia, 326
Pragmar, 331
Prazalanz, legend of, 26
Predajo, 342
Predazzo, 377; road to Treviso, 377
Primiero, 377, 398, 400
Primolano, 32, 400
Puss in Boots, 431-2
Pusterthal, 286

Q UARAZZA, 396
Quirinus, St., 330

R ABBI VAL DI, 363, 438
Ragatz, 23
Raggal, 25
Rankweil, 14-6
Ratfeld, 66
Rattenberg, 66
Reif, 406
Reit, 324
Rettenberg, 152-3, 204-5
Rettengschöss, 126
Reutte, 34, 35
Revo, 365
Rhaetius, 347
Rheiche Spitze, 129

SAD

Rheinthalersee, 127
Rhine, 13, 17, 20, 42, 52
Riddles, 424, 440
Riedenberg girls' college, 20
Riedl, Sebastian, 84
Ring, 409 note, 425, 428
Rinn, 151, 212 et seq.
Riva, 400, 406-7
Rivoli, 341
Robin Goodfellow, 113-4
Robler, legends of, 95, 182-4, 221-2
Rochetta Pass, 358-9
— in Val Sugana, 393, 395
Roesla Sandor, 55
Rofnerhof, 239, 240
Rofreit, 342
Roman remains, xix-xx, 307, 395, 407, &c.
Romedius, St., 165
Romediusschlucht, hermitage in the, 368
Romediusthal, 368
Rosanna, 30
Rosemary, 103
Rosenheim, 45
Rost, 297
Rothenbrunn, 330
Rothholz, 66
Rottenburg, 66 et seq., 286
Rottenburger family, 66 et seq., 83
Roveredo, 342, 344
Rudolf of Hapsburg, 253, 271
— II. of Tirol, 273, 276
Rufreddo torrent, 368
Ruined castles of Tirol, merits of, xi, 189-90
Rum, 167

S ÄBEN, 356 note
Sadole, 398

INDEX.

SAE

Sætere, 4
Snigesbach, 330
Säligen, Fräulein, 413, 428
Salvan, 408 et seq. 413, 420
Salvanel, 412
Salzberg and saltworks, the, 155-7
Sandbichler, the Bible commentator, 124
Sandwirth, the, 150
Sanzeno, 367
Sarcathal, 400
Satteins, 16
Saturn, temples to, in Tirol. 356, 360, 372-4, 387
Sayings, popular, of Wälsch-Tirol, 439-40
Scharnitz, 324, 325-6
Schattenberg, 26, 43
Scheibenschiessen, 94
Schildhof, the, 241
Schlitters, 82
Schloss Junk, 344
Schmalkald League, 21, 247
Schneeburg, 161
Schnodahüpfl. 98, 346
Schrofenstein, 31
Schruns, 28
Schwangau. 34
Schwarzen Felder, the, 360
Schwarzensteingrund, 91
Schwatz, 145, 168, 320
— its frescoed houses, 170
— inns, 170-2, 186-8
— history, 173 et seq.
— mines, 173 et seq.
— shops, 177
— parish church, 178 et seq.
Schwefelloch, the, 315
Scotch missionaries to Tirol, 14 et seq.

SCU

Sculpture, curious early, 25, 30. 31, 65, 66, 216, 243, 249 et seq.
Sechsmilionen-Brücke, 358
Seefeld. 194 et seq., 270, 324-5
Seidenbaum, the, 360
Scirens, 126
Sellathal, 393, 396
Selrain, 330, 332, 338
Senale, 366
Sendersbach und thal, 333, 338
Serlesspitz, 339
Serravalle, 344
Sette Comuni, 393, 396
Seven Years' War, 289
Siechin, die fromme, 258
Sigismund, St., 325, 331
Sigismund, Emperor, 153, 240
— (the Monied), 146, 150, 243, 252, 319-20, 337
Sigmund Frank, 284
Sigmundsburg, 33. 244
Sigmundsegg, 244
Sigmundsfried, 244
Sigmundslust, 146, 244
Sigyn, 3
Silberthal, the, 24
Silk, 342, 344-5, 360, 386, 396
Silver chapel, the, 254, 269-70
Simeon of Trent, St. 215, 349
Simpleton stories, 432-7
Sisinius, Martyrius, and Alexander, SS., 360, 367-374
Sistrans, 219; legend of, 221-2, 380 and note
Skolastica, 132
Skulls kept unburied, 30, 65
Slavini di San Marco, 342, 343

SUN

Sleeper, legend of the, 388-90
Snail, story of a horrible, 435-6
Solstein, 315, 324
Sondergrundthal, 88
Sonnenberg in Vorarlberg. 26
Sonntag, 25
Spaur, 157
— Maggiore, 359
Speckbacher, 55, 150-1, 156, 300
Sphinx, 424
Spider, 268
Spiel-joch, 132
Spinges, das Mädchen von, 297
Stainer, Jacob, 160-1
Stalactites, 396
Stallenthal, the, 137 et seq.
Stams, 195, 247
Stans, 136
Stanzerthal, 29-30
Starkelberg, 206
Starkenberg, 96
Stase-Sattel, 46-8
Steeples, legends of, 65 and note
Steinach, 91
Steinberg, 127
Stelvio, 31
Sternsingen, 100
Sterzing. 59, 160, 297
Stillupethal, 88
Stones, supernatural, 15, 212, 311, 404, &c.
Strass, 82 et seq.
Strigno, 396-7
Stubaythal, 147, 338, 339
Stuibfall, the, 36
Sugana, Val, 341
Sulz, 315
Sulzberg, 358
Sundays and festivals, legends about, 71, 196-7, 380-1

452　INDEX.

SWA
Swallow in Tirolean mythology, 197, 439
Swinburne, General, 326–7
Swiss embroidery, 21–2, 23, 42

TAILOR, VALIANT, 358 and *note*
Tamina-schlund, the, 23
Tapers, superstition about, 104
Tarpeian rock, the, parallel, 24 *note* 2
Tarquini,[1] Cardinal, on Etruscan remains in Tirol, 372–4 and *note*
Tavola Clesiana, 361–3
Taur, 153, 164–6
Taylor, Rev. Isaac, xvi, 79, 131, 373, &c.
Telfs, 325
Templars, 406
Terfens, 146
Teufelspalast, 350
Teufelsteg, the, 88
Tezze, Le, 32, 399, 400
Theophrastus Paracelsus, 265 *et seq.*
Thierberg, 59, 112, 147
Thiergarten, 126
Thirty Years' War, 275, 282, 354
Thor, 6
Thun family, 368–9
Thürl, 164
Thurnegg, 80
Thyrsus, 231–3
Tiefenbach, Friedrich, 277
Tione, 405

TIR
Tirol, Schloss, xix, xx, 24, 41, 241
Titian, 401–2
Toads in Tirolean mythology, 26, 157–8, 181
Tobacco, 176, 307, 345, 407
Tobel, 360
Toblino, 401–2
Todsündenmarterle, 222
Tonale, 365
Trüchnabächle. 26
Traitors, legends of, 24, 379
Tratzberg, 80–2, 201
Treasure legends, 26, 324, 343–4, 434. *See* also Mining Legends
Tree-hermits, 16 [2]
Tregiovo, 366
Trent, 214–5, 248, 286, 345, 439
Trent Festa of St. Vigilius, 346
— history, 346–7
— churches, 348
— legend of the crucifix, 348–9
— St. Simeon of, 349
— clubs, 350
— Museum, 350, 372 *note*
— private palaces, 350
— Teufelspalast, 350
— suburbs, 354 *et seq.*
— Council of, 246, 348–9
— — dispersed witches, 419
Trisanna, 30
Trisenega, 359
Troi, 202

VAL
Tschirgants, 33
Turlulù, 432–4
Turquoise in Tirol, 200
Tuscan customs and traditions, 417 *note*
Tuxerthall. *See* Duxerthal

UDALRIC, St., 397
Ulrich, St., 381
Unmarried, legend of the, 184–6
Unnutzjoch, 132
Unter Perfuss, 329
Urban, St., 439

VADUZ, 23
Val Arsa, 420, 438
— Avisio, 311
— Bona, 400
— Breguzzo, 400
— Camonica, 365
— di Cembra, 374
— Daone, 400
— d'Inferno, 389
— di Fiemme, 374
— Génova, 405
— Lagarina, 341 *et seq.*
— di Ledro, 400
— di Non, 341, 358 *et seq.*, 420, 421
— del Orco. 388
— di Pejo, 364
— di Rabbi, 363, 438
— Rendena, 400, 402 *et seq.*
— Sella, 393, 395–6
— di Sole, 341, 358, 363 *et seq.*
— Sugana, 341, 382 *et seq.*

---

[1] It is noteworthy that so prominent an enquirer into Etruscan antiquities should bear a patronymic so connected with Etruria as Tarquini.

[2] In Abbé Dubois' introduction to his translation of the Pantcha Tantra, is a story called 'La fille d'un roi changé en garçon,' in which mention is made of a Brahman hermit who fixed his residence in a hollow tree.

INDEX.

VAL

Val Vermiglio, 364
Valduna, 16, 24–5
Valentin, San, hermitage, 387
Vampires, 136
Veitskapf, St., 13
Vela, Acqua della, 404
Vendôme in Tirol, 286 *et seq.*
Ventthal, 428 *note*
Verda, 382
Verdes torrent, 368
Verruca, 356
Vezzena, 392
Viecht, 80, 133 *et seq.*, 143 *et seq.*
Vierzehn Nothhelfer, the, 36–8
Views, striking, in Tirol, 13, 17, 20, 21, 33–4, 133, 202, 268–9, 330, 355–6, 365–6, 376, 386, 392, 397, 401, 407, &c.
Vigilius, St., 165, 346–8, 360–2, 367, 402 *et seq.*, 440
Vigo, 379–81
Villerspitz, Hohe, 330
Vintschgau, the, 99, *note* 410
Visiaun, 359
Vitus, St., legend of, 38
Volandseck, 147
Voldepp, 126
Volders, 150, 202, 206, 207, 210
Völlenberg, 337
Vöis, 334
Vomp, 145–6

VOR

Vorarlberger-gheist, 51–2

WAGNERWAND, the, 315
Waidburg, 337
Walchen, 205
Waldauf, Florian, 151–2, 204
Waldrast, 147, 338–9
Wälschmetz, 358
Wälsch Michel, 358
Wälsch-Tirol, Italian views upon. See Italians
Wallachian legend, 404, 405
Wallenstein's conversion, 265
Wallgau, xv. xvii, 24
Waterfalls, 32, 36, 86, 88, 330, 333, 356, 366, 397
Wattens, 202, 206
Weer, 202, 204
Weierburg, 313–4
Weight-offering, Buddhist, 335
Welser, Philippine, 259 *et seq.*, 308
Wendelgard and Ulrich, legend of, 18–19
Wiesburg, 30
Wiesing, 126
Wildenfeld, the, 134
Wild men, 35
Wildschönau, the, 111 *et seq.*

ZUL

Wilten, 230 *et seq.*, 278, 293
Wishes, 47, 425 *et seq.*
Witches' Sabbath, 365, 415
— bridge, 407
— dispersed by Council of Trent, 419
Wodan, 4, 6, 119
Wolf-boy mythology, 420 *et seq.*
Wolfurth, 20
Wolves in Tirol, 31
Wörgl, 132
Worms, 31
Wrestling, 95

ZABERLE, 396
Zambelli, Palazzo, 350
Zamsergrund, the, 91
Zell, near Kufstein, 63
Zell in Zillerthal, 79
Zemgrund, the, 91
Zemmerferner, 88
Zeno, St., 367–9
Ziano, 398
Zillerthal, the, 79 *et seq.*, 92 *et seq.*, 202
Zimmerthal, 374
Zips, Oswald, 322
Zircinalpe, 126
Zirl, 256, 314, 323–5, 329
Zocco di Natale, 438
Zuam dall' Orso, 420 *et seq.*
Zugspitzwand, 34
Zulian, San, 404

_{*} I have endeavoured to make the nomenclature as consistent as possible throughout, but great diversity both of pronunciation and spelling prevails in Tirol itself.

LONDON: PRINTED BY
SPOTTISWOODE AND CO., NEW-STREET SQUARE
AND PARLIAMENT STREET

www.ingramcontent.com/pod-product-compliance
Lightning Source LLC
Chambersburg PA
CBHW021426300426
44114CB00010B/672